Artificial Et

Artificial Ethology

Owen Holland
University of the West of England

and

David McFarland
University of Oxford

UNIVERSITY PRESS

OXFORD
UNIVERSITY PRESS

Great Clarendon Street, Oxford OX2 6DP
Oxford University Press is a department of the University of Oxford.
It furthers the University's objective of excellence in research, scholarship,
and education by publishing worldwide in
Oxford New York
Athens Auckland Bangkok Bogotá Buenos Aires Calcutta
Cape Town Chennai Dar es Salaam Delhi Florence Hong Kong Istanbul
Karachi Kuala Lumpur Madrid Melbourne Mexico City Mumbai
Nairobi Paris São Paulo Shanghai Singapore Taipei Tokyo Toronto Warsaw
with associated companies in Berlin Ibadan

Oxford is a registered trade mark of Oxford University Press
in the UK and in certain other countries

Published in the United States
by Oxford University Press Inc., New York

© Oxford University Press 2001

The moral rights of the author have been asserted
Database right Oxford University Press (maker)

First published 2001

All rights reserved. No part of this publication may be reproduced,
stored in a retrieval system, or transmitted, in any form or by any means,
without the prior permission in writing of Oxford University Press,
or as expressly permitted by law, or under terms agreed with the appropriate
reprographics rights organization. Enquiries concerning reproduction
outside the scope of the above should be sent to the Rights Department,
Oxford University Press, at the address above

You must not circulate this book in any other binding or cover
and you must impose this same condition on any acquirer

British Library Cataloguing in Publication Data
Data available

Library of Congress Cataloging in Publication Data
Data available

ISBN 0-19-851057-8 (Pbk)

Typeset by The Author
Printed in Great Britain
on acid-free paper by
Biddles Ltd, Guildford & King's Lynn

Preface

In recent years there has been rapid growth in the application of robotics and AI techniques to problems of animal behaviour. The range of activities includes the study of the behaviour of real and simulated robots, the use of robots to gather data from animals or the environment, and the investigation of the performance of animal-like robots in real-life situations. Many ethologists are not fully aware of this new approach to their subject. One aim of this book is to inform ethologists about how work on robots can benefit the study of animal behaviour. All the contributors to this book were trained originally in some form of life science, and all have first hand experience of real robots. All are able to review the advantages that can be gained from the robotics approach.

A second aim of this book is to inform those readers trained in the physical sciences about the ways in which robotics can be seen as a tool in biological research. For this reason the main text is pitched at an elementary biological level, to bring the physical scientists into the picture; while a series of case studies presents biological detail, without too much technical robotics that might not be understandable to a biology student.

This book is based upon the proceedings of the Artificial Ethology Workshop, which was held at Puerto Chico, Playa Blanca de Yaiza, Lanzerote in July 1998. The authors would like to thank the sponsors: Oxford University Press, and The Faculty of Engineering, University of the West of England; the organizers: Animalia and Park Offices, Playa Blanca; and all participants at the International Workshop in Artificial Ethology.

Owen Holland and David McFarland

Contents

Preface v

Chapter 1. History of models in ethology 1

Chapter 2. The evolution of animal-like mobile robots 14

Chapter 3. Sensory processes and orientation 42
    Case Study 1: How robotic lobsters locate odour sources in turbulent water
    Frank Grasso 47
    Case study 2: Robotic experiments on cricket phonotaxis
    Barbara Webb 59
    Case study 3: Gathering and sorting in insects and robots
    Owen Holland 77

Chapter 4. Motor co-ordination 93
    Case study 4: How frogs groom
    Simon Giszter 97
    Case study 5: Robotic experiments on insect walking
    Holk Cruse 122
    Case study 6: Building a robotic lobster
    Jospeh Ayers 139

Chapter 5. Motivation and learning 156
    Case study 7: Neural nets and robots based upon classical ethology
    Janet Halperin 162
    Case study 8: Robotic experiments on rat instrumental learning
    Emmet Spier 189
    Case study 9: Robotic experiments on complexity and cognition
    Brendan McGonigle 210

Chapter 6. Why robots? 225

Bibliography 236

Index 257

Contributors

Joseph Ayers Marine Science Center, Northeastern University, East Point, Nahant, MA 01908, USA

Holk Cruse Department of Biological Cybernetics, University of Bielefeld, Postfach 100131, D-33501 Bielefeld, Germany

Simon Giszter Kinesiology and Life Sciences Consortium, Pennsylvania State University, 266 Recreation Building, University Park, PA 16802, USA

Frank Grasso Boston University Marine Program, Marine Biological Laboratory, Woods Hole, MA 02543, USA

Janet Halperin Department of Animal Sciences, University of Maryland, MD 20742-556, USA

Brendan McGonigle Laboratory for Intelligent Systems and Cognitive Neuroscience, Department of Psychology, Edinburgh University, Edinburgh EH8 9JZ, UK

Emmet Spier School of Cognitive and Computing Sciences, Sussex University, Brighton BN1 9QH, UK

Barbara Webb Department of Psychology, University of Stirling, Stirling, FK9 4LA, UK

Chapter 1

History of models in ethology

Models and analogies have been a feature of ethology since its early days. Konrad Lorenz (1950) likened the motivational system of an animal to the simple hydraulic system illustrated in Figure 1.1. In this model, motivational 'energy', represented by the liquid in the reservoir, builds up in the absence of an opportunity to perform the relevant behaviour, and 'discharges' when the relevant external stimuli (represented by the weight on the pan) 'release' the behaviour. Apart from the inappropriateness of the concept of energy (Hinde 1960; McFarland 1971) this model offers a good analogy, at a general level, of the interaction of internal and external factors in animal motivation. However, ethological modelling of a much more precise nature considerably predates Lorenz's attempts.

Erich von Holst, a contemporary of Lorenz, pioneered the subject of behavioural physiology, and was the first to use modelling techniques systematically in behavioural research. In 1939 he published a long paper on relative co-ordination, based upon his research into swimming co-ordination in fish. Von Holst (1939) discovered that, in teleost fish, the fins beat rhythmically, but they have some degree of independence. Under certain circumstances they beat at different frequencies. The rhythms of different fins influence each other, a feature known as relative co-ordination. Sometimes the rhythm of one fin attracts and dominates that of another (called the magnet effect) so that they fall into step. In other cases the amplitudes of fin movements summate, so that the movements become smaller when the fins are out of step with each other, and more extensive when they are in step. In this paper von Holst employed a sophisticated mechanical model to illustrate his discoveries (see Figure 1.2).

The publication of Norbert Wiener's *Cybernetics* in 1948 provided a unifying theory and nomenclature for ideas that were nascent in biology. The application of cybernetic ideas in biology stems from the recognition that analogies can exist between the behaviour of physical and biological systems. Analogous laws of nature had long been recognized in the physical sciences. Two systems are said to be analogous when their behaviour, defined by a mathematical equation, is identical. The equations describing the behaviour of such systems are thus

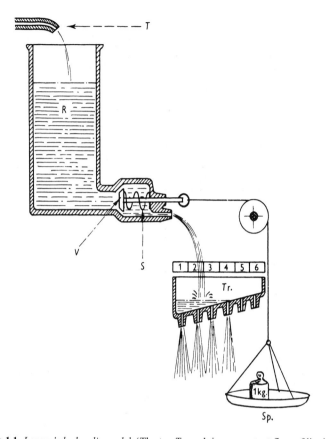

Figure 1.1. *Lorenz's hydraulic model.* 'The tap T supplying a constant flow of liquid represents the endogenous production of action-specific energy; the liquid accumulated in the reservoir R represents the amount of this energy which is at the disposal of the organism at a given moment, the elevation attained by its upper level corresponds, at an inverse ratio, to the momentary threshold of the reaction. The cone valve V represents the releasing mechanism, the inhibitory function of the higher centres being symbolized by the spring S. The scale-pan Sp which is connected with the valve-shaft by a string acting over a pulley represents the perceptual sector of the releasing mechanism, the weight applied corresponds to the impinging stimulation. This arrangement is a good symbol of how the internal accumulation of action-specific energy and the external stimulation are both acting in the same direction, both tending to open the valve.' (Lorenz 1950). The outflow from the reservoir represents the motor activity of the behaviour, the intensity of which can be measured by the distance the jet reaches. The trough Tr symbolizes the recruitment of different components of the behaviour pattern as the intensity increases. The intensity of the observed behaviour is a joint function of the strengths of the internal and external causal factors.

independent of the hardware of the system(s). This fundamental realization opened the way for the development of a control systems theory applicable to both physical and biological systems, and von Holst and his students were quick to exploit this new thinking and introduce it into ethology.

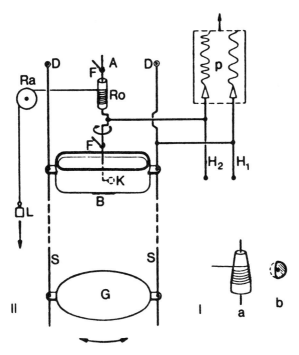

Figure 1.2. *Von Holst's mechanical model of swimming co-ordination.* Two bars, freely suspended from points D, are loaded with a weight G, and carry a reservoir B containing a viscous fluid (syrup). The pendulum system is powered by the weight L which is suspended over the pulley Ra by a thread which is wound round the cylinder Ro. Oscillation of the pendulum moves the recording lever H_1. The axle A rotates around a fixed point F, and its lower end operates the recording lever H_2, and bears an eccentric sphere K which dips into the reservoir B.

Initially the insights gained from this new approach were qualitative. For example, von Holst and Mittelstaedt (1950) extended and generalized the outflow theory of Helmholtz (1867). Any sophisticated orientation system must be able to distinguish between stimulation from the outside world and stimulation caused by the animal. In the case of human vision, for example, movement of objects in the outside world causes movement of the image on the retina, which we perceive. Voluntary movement of the eyes, however, also produces movement of the image on the retina, yet we do not perceive such movements. Somehow the brain distinguishes between the movement of the retinal image that is independent of the animal and movement that is due to movement of the eyeball.

Two theories have been proposed to account for this phenomenon: the outflow theory and the inflow theory. The outflow theory, originally due to Hermann von Helmholtz (1867), maintains that instructions to the eye muscles to move the eyeball are accompanied by parallel signals to a comparator in the brain. Here they are compared with the incoming visual signals, as shown in

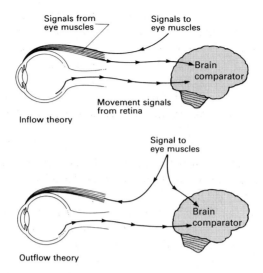

Figure 1.3. *The inflow and outflow theories of eyeball movement.*

Figure 1.3. Inflow theory, due to Charles Sherrington (1918), maintains that receptors in the extraocular muscles send messages to the brain comparator whenever the eyes are moved. In both theories, the brain comparator assesses the two incoming signals and determines whether the visual signals correspond to the movement that would be expected on the basis of the other signal. If the two signals do not correspond, then some of the movement must have been due to outside causes.

The outflow theory was extended and generalized by Eric von Holst and Horst Mittelstaedt (1950) (see also von Holst 1954). According to their reafference principle, the brain distinguishes between exafferent stimulation (stimulation that results solely from factors outside the animal) and reafferent stimulation (that which occurs as a result of the animal's bodily movements). Motor

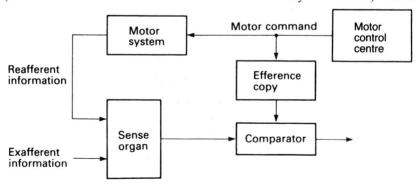

Figure 1.4. *Outline of a basic reafference system.*

commands not only cause patterns of muscular movement but also produce a neural copy (the 'efference' copy) that corresponds to the sensory input that could be expected on the basis of the animal's behaviour. The brain then makes a comparison between the efference copy and the incoming sensory information (see Figure 1.4). All reafferent information should be cancelled by the efference copy so that the output of the comparator will be zero. Exafferent information will not be cancelled, however, and will be passed on by the comparator to another part of the brain.

Von Holst and Mittelstaedt (1950) showed that the fly *Eristalis*, when placed inside a cylinder painted with vertical stripes, shows a typical optomotor reflex, that is, turning in the direction of the stripes when the cylinder is rotated (Figure 1.5a). Such reflexes do not occur when the fly moves of its own accord, although the visual stimulation is similar. When the head of the fly is rotated experimentally through 180° (see Figure 1.5b), the optomotor reflex is reversed as expected. However, when the fly attempts to move of its own accord, it goes into a spin and its movements appear to be self-exciting. These results can be explained in terms of reafference theory. Normally, the output of the comparator determines bodily movement, and when the fly moves of its own accord, the output is zero and no movement occurs. The optomotor apparatus provides exafferent stimulation that is not cancelled by an efference copy, and the fly responds in a reflex manner. When the head of the fly is reversed, exafferent stimulation has the same effect as before but in the opposite direction. In the case of reafferent stimulation, however, the perceived movement is reversed in

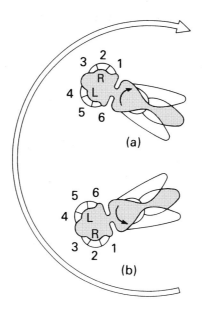

Figure 1.5. *The optomotor response of flies.* (a) Before head rotation. (b) After head rotation.

sign, and instead of being subtracted from the efference copy, it is added to it. The result is that the comparator now has a magnified output, whereas normally it would be reduced to zero. The more the animal responds, the greater the reafferent stimulation and the greater the amplification of the animal's response. Consequently, the fly tends to spin faster and faster.

Thus von Holst and Mittelstaedt formalized a previously vague theory and were able to test it experimentally. Their reafference principle is important not only with respect to vision but also in the control of limb position, posture, etc. For example, we can tell the difference between the arm movements involved in shaking the branch of a tree and those, which may be identical, produced while holding passively onto a branch that is being moved by the wind. However, the reafference model was never a fully quantitative model. Like most of the cybernetic models of the fifties, it did not make very precise predictions.

For this to happen, it was necessary to introduce time as an explicit variable. In behavioural research, this began to happen in the 1960s. Examples are Stark's (1959) and Clynes' (1961) studies of pupillary light reflexes, McFarland's (1965) study of drinking in doves, and Horridge's (1966) study of the optokinetic response of crabs (see McFarland 1971 for further details).

By the late 1960s, various methods of 'black box analysis' were being used to try to identify a transfer function for the whole animal (with respect to specific inputs and outputs), or to try to identify the major state variables. These included frequency analysis (McFarland and Budgell 1970; Oatley and Toates 1971), stochastic identification techniques (McFarland and Lloyd 1971; McFarland and Rolls 1972; Lloyd 1974), and stochastic analysis of behaviour (Delius 1969; Heiligenberg 1969, 1974; for discussion see McFarland 1971, 1974a; Toates 1975).

As an example of the problems inherent in these approaches, let us look at the study by McFarland and Rolls (1971). The study concerned the question of whether there is a central inhibition by thirst on feeding. It had been known for years that thirsty animals eat less than normal. Most water-deprived animals take action to conserve water, and cutting down on food is an obvious way to do this. Eating less makes the animal hungry, and the evidence shows that thirsty animals will work for food that they do not eat. It appears, therefore, that there is an inhibition by thirst on eating per se. To test this hypothesis, McFarland and Rolls (1971) trained rats to work for food and water in a Skinner box, and then tested them when hungry, with access to food only. While the rats were working for food they were given a single intracranial injection of angiotensin, a known thirst stimulant. In all cases the rate of feeding was depressed and the time-course of the depression was recorded (see Figure 1.6). To obtain a response of this magnitude, a large dose of angiotensin is required, and it could be argued that this situation was unnatural. McFarland and Rolls (1971) carried out another experiment in which the angiotensin injections were controlled by a computer, and delivered in a pseudo-random binary sequence (PRBS). The syringe was driven at a constant rate when the input was on, and at a zero rate when the input was off. The total dose delivered during the test session was comparable with the first experiment, but the individual doses were extremely small. The computer was used to record the rat's feeding behaviour and perform

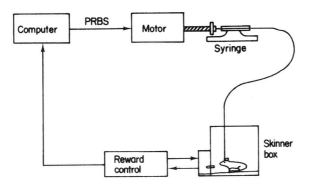

(a) Computer-controlled chemical stimulation of the brain. Intracranial injections are administered to a rat in a Skinner box, in accordance with a PRBS signal. The effect of the injections on the rat's bar-pressing performance is recorded by the computer.

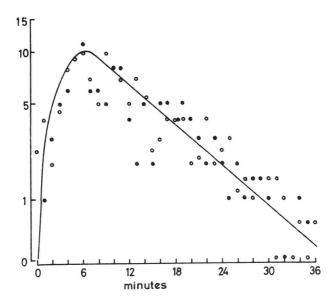

(b) The effect of intracranial angiotensin on operant feeding. The open and closed circles represent data from two rats showing depression of feeding following a single injection. The ordinate scale denotes a measure of the magnitude of the depression of feeding. The solid line indicates the calculated impulse response.

Figure 1.6. *Determination of a behavioural response by a stochastic identification technique.*

a cross-correlation with the PRBS input. This computation yields the impulse response of the system (see McFarland 1971 for details). The impulse response gives the effect on the feeding system of a hypothetical impulse injection of angiotensin, and this is illustrated in Figure 1.6.

Another useful approach is stochastic analysis of observed data. For example, as a result of his stochastic analysis of aggressive behaviour of the cichlid fish *Haplochromis burtoni*, Heiligenberg (1974) arrived at the control model illustrated in Figure 1.7. Various types of stochastic analysis have been employed in the formulation of motivational models, including multivariate analysis of variance, Markov chain analysis (Metz 1974), and stochastic identification techniques (Lloyd 1974) like that described above.

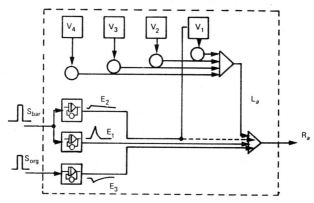

Figure 1.7. *Analogue computer model representing readiness to attack (R_a) in a cichlid fish. R is the sum of an internal variable (L) and of processes (E) which are triggered by an external stimulus S. $V_1 - V_4$ are the internal states obtained by stochastic analysis.* (After Heiligenberg 1974.)

While these approaches seemed successful at the time, and led to a spate of 'how to do it' books (e.g. McFarland 1971; Toates 1975, 1980), their application was limited to particular systems, the inputs and outputs of which could be readily measured. To handle whole animal behaviour from a systems analysis viewpoint, methods appropriate to multi-input-output systems are required. While classical control systems theory is ideal for analysing the dynamics of relatively simple linear systems, it is not well suited to the multi-input/non-linear output situation that characterizes the real animal. The fundamental non-linearity of animal behaviour arises from the fact that, in the unconstrained animal, a number of parallel systems compete for behavioural expression. The animal switches from one type of behaviour to another, often in complex ways (McFarland 1974*b*; Ludlow 1980). As an example, let us consider the courtship behaviour of the smooth newt (*Triturus vulgaris*), as described by Halliday (1974). The courtship takes place in the spring when the newts return to ponds after hibernating on land. During this aquatic phase, the

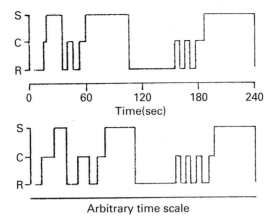

Figure 1.8. Comparison of the mean durations of various courtship activities of 22 newts (above) with that of the NEWTSEX model (below). R = retreat display, C = creep, S = spermatophore transfer. (After Houston *et al.* 1977.)

newts must swim up to the pond surface to breathe, so they must hold their breath during the underwater courtship. Fertilization is internal and is achieved by the male depositing a spermatophore on the floor of the pond. The female then walks over the spermatophore and, in a successful courtship, picks it up with her cloaca. The model we wish to discuss is only concerned with the two activities of the male that precede this transfer and which Halliday calls retreat display and creep. During retreat display, the male backs away from the advancing female while performing various movements with his tail. This phase ends when the male turns away from the female and creeps along the floor of the pond. The female follows him and when he quivers she touches his tail with her snout. This appears to be the stimulus for him to deposit a spermatophore. Halliday (1974) found that, on the first of several spermatophore depositions, males tended to complete this prototypical sequence (retreat display – creep – spermatophore transfer) without interruption. On the last spermatophore, however, they often alternated between retreat display and creep before depositing a spermatophore (see Figure 1.8). Furthermore, the display rate is correlated with the number of spermatophores still to be deposited. This number, which is known only at the end of the courtship, will be referred to as S (S usually ranges from 1 to 3).

Houston *et al.* (1977) developed a control model, called NEWTSEX IV, that could mimic the newt's behaviour. The model (see Figure 1.9) is based on a variable called *Hope*, which can be thought of as the male's assessment of the sexual state of the female. The initial value of *Hope* depends on S, and its rate of change is determined by the consequences of the male's behaviour, represented by feedback loops with parameters k_2 and k_4. During retreat display, when the male can see the female approaching him (represented by F in Figure 1.9), *Hope*

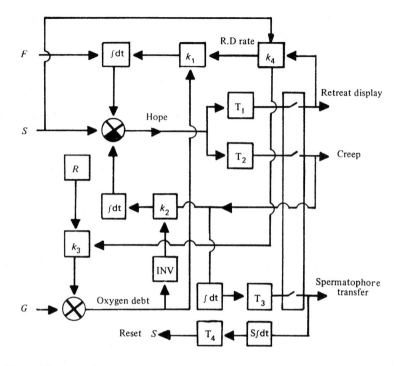

Figure 1.9. *A control model for the courtship of the smooth newt.* S = initial amount of sperm, F = initial sexual state of the female, G = initial amount of *Oxygen need*, R = function representing the increase of *Oxygen need* with time, $T_1 - T_4$ = thresholds, $k_1 - k_4$ = modifiable parameters. (After McFarland 1977, modified from Houston *et al.* 1977.)

rises (i.e. the feedback is positive). When the male turns away to creep he can no longer see the female, so it is assumed that *Hope* falls because of negative feedback. The transition from retreat display to creep is determined by a threshold called T_2. Once creep begins, *Hope* starts to fall, so that if T_2 were also the threshold for the transition from creep to retreat display, the model would 'dither'. This is a common problem in modelling animal behaviour. In NEWTSEX IV this difficulty is overcome by having a lower threshold T_1, which determines the transition from creep to retreat display. This means, however, that between T_1 and T_2 the value of *Hope* does not suffice to determine whether retreat display or creep occurs: the direction in which *Hope* is changing must also be known.

To explain the influence of the number of spermatophores to be deposited on the nature of the courtship sequence, it is assumed that the rate at which *Hope* falls is inversely proportional to S. NEWTSEX IV assumes a fixed duration of creep after which the spermatophore transfer begins. This means that if *Hope* falls slowly, as it does when $S = 3$, there is time for creep to be completed before T_1 is reached. When $S = 1$ or 2, *Hope* may get to the threshold before there has been enough time for creep, and so the model switches back to retreat display. As it stands, if this model fails to complete creep at the first attempt, it will alternate between retreat display and creep indefinitely. To enable it to progress to the transfer phase, we introduce a variable called *Oxygen need* that represents the internal availability of oxygen. We are not in a position to make any precise statement about the physiological basis of this variable, but we assume that it increases with time under water, and that the rate of increase is proportional to the male newt's rate of activity. This increase is represented by the ramp function R that is influenced by the rate at which retreat display is performed. The size of the gulp of air that the newt takes at the surface is modelled by the value of G, which is subtracted from the ramp function R.

Adding the *Oxygen need* variable to the model enables the alternation between retreat display and creep to be discontinued because it is assumed that as *Oxygen need* increases, *Hope* falls less rapidly during creep. As can be seen from Figure 1.8, the output of the resulting model is similar to the behaviour of the male newt. Some implications of the model were tested experimentally by Halliday (1977). He changed the oxygen content of the atmosphere above the water in the newt's tank. This should change the value of G, and hence the timing of the courtship sequence. If it is assumed that the value of G is bigger than usual when the atmosphere is enriched with oxygen, and smaller than usual when extra nitrogen is added, then the model predicts that the courtship will be slower than normal in the former case, and faster in the latter. This prediction was supported by Halliday's results, as shown in Figure 1.10.

This example illustrates an important problem associated with modelling animal behaviour in terms of control systems theory. An animal does one type of behaviour (activity) at a time, so the animal's motivational state can never be completely represented by the behaviour. The state of the animal involves many

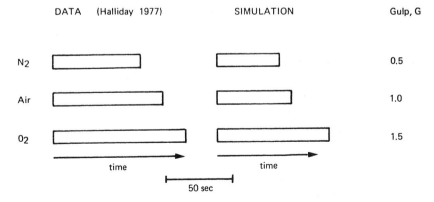

Figure 1.10. *Comparison of data from real newts and control model.* Duration of newt courtship from the start of retreat display to the deposition of the first of three spermatophores, under three conditions of water and surface gases: high nitrogen content, normal air, and high oxygen content. (From McFarland and *H*ouston 1981.)

state variables, only some of which are relevant to the ongoing behaviour. The other state variables may be relevant to future behaviour, but the inability of the observer to monitor these variables causes characteristic observability problems.

The concept of observability was developed by Kalman in the early sixties, when he was investigating the relationship between input/output (transfer function) and state variable descriptions of linear dynamical systems. He established that the input/output relationship determines only that part of the system that is completely controllable and completely observable. Complete controllability means that any state can be moved to any other state during a finite time interval by a suitable input. Complete observability means that the state can be determined uniquely from the input and output over a finite interval. Furthermore, this part of the system has the smallest dimensionality of all realizations of the input/output behaviour (Kalman 1963). Questions of observability and controllability are of profound importance in modelling animal behaviour. They are discussed at length by Houston *et al.* (1977) in relation to NEWTSEX IV. McFarland and Sibly (1972) discussed the problem of observability from the animal's viewpoint. The animal may have an internal goal (i.e. a desirable internal state) which the environment does not enable it to obtain. If it makes a 'best' attempt at reaching the goal, it may behave in exactly the same way as an animal that can reach its goal. In their words, 'Observable commands are in one-to-one correspondence with their optimal consequences, whereas the relationship between unobservable commands and their consequences is equivocal.'

The intractability of modelling animal behaviour in the face of observability problems and problems to do with the scaling of internal variables (Houston and McFarland 1976) led to a new approach to modelling. Instead of trying to model directly the decision-making mechanism of the animal, we can turn the question the other way around. Given the state of the animal, and the information provided to the brain about that state, what decision ought the animal to make to achieve certain objectives? The difference between the two approaches is illustrated in Figure 1.11.

Questions about what animals ought to do in a given set of circumstances are essentially functional questions, rather than questions about causes. If we ask what makes an animal eat rather than drink in a given situation, then we are asking a question about the causal factors for eating and drinking. Once we start asking about causal relationships between systems, rather than within systems, then we come up against problems about the equivalence of the various factors that belong to different systems. In modelling such a situation, the assignment of equivalencies is essentially arbitrary, without reference to some outside criterion. Such criteria can only be provided by a functional model.

During the 1970s there was a gradual realization that some kind of functional modelling was necessary if progress was to be made in understanding animal decision making. For example, such an approach is required in considering the

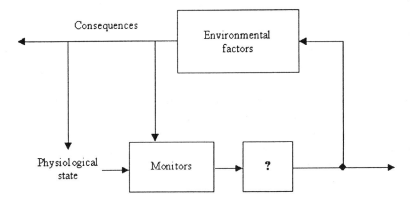

Figure 1.11. *Optimal control model.* In an optimal control model, the question (?), of what the control laws should be to obtain a certain objective, is essentially a functional question. In the case of animal behaviour, the question is what decision rules (?) should be employed to produce the 'best' consequences.

courtship of the male newt, described above (McFarland 1977). Functional modelling inevitably leads to optimality questions: what is the best way for the animal to behave in a given situation, given that it should trade-off amongst a variety of causal pressures? Posing this question (e.g. McFarland and Houston 1981) led to a whole new approach to modelling animal behaviour from a functional viewpoint, and to the demise of purely causal modelling.

Functional modelling has taken various forms, because the basic questions can be looked at from an economic (e.g. Allison 1979), a sociobiological (e.g. Trivers 1985) or a behavioural ecology (e.g. Stephens and Krebs 1986) viewpoint. New techniques, such as state-space analysis, dynamic programming, and stochastic analysis have led to considerable advances in our understanding of animal decision making, but they are beyond the scope of this review, because they do not answer the fundamental question: What is the actual mechanism of decision making in a given case? As we shall see from case studies within this volume, the introduction of animal-like robots, as a new type of scientific tool, provides fresh hope that this kind of question can be revived, after so long in the wilderness.

Chapter 2

The evolution of animal-like mobile robots

As we have seen, early ethological models were often presented in terms of the familiar characteristics of contemporary physical systems, as in von Holst's (1939) mechanical model, or Lorenz's (1950) hydraulic model. As new technologies developed, models expressed in terms of these technologies appeared; for example, Heiligenberg's (1974) model of the determinants of attack behaviour in cichlid fish used the symbolic language of analogue computing. However, perhaps the greatest influence on modelling in the last few decades has been the digital computer. This provided both a general model of information processing, and a range of techniques for describing and analysing information-processing systems. The scientific community soon became familiar with these ideas, and the computer became the analogy of choice: concepts such as software, hardware, flow diagrams, programs, data, processing, and memory storage became pervasive. Models could then be presented in quite abstract terms, with a computer simulation serving as an adequate investigative, illustrative, and explanatory tool.

Many models in ethology are designed to show how internal factors combine with the input from external factors to determine the choice of some observable behaviour. When that behaviour is executed, it may change the input from external factors, either by changing the environment, or by changing the effect of the environment on the animal – for example, by changing the animal's orientation to or distance from some environmental feature; this change in input may affect subsequent behavioural choices. Some models are simple enough for the link between a given behaviour and the input from external factors to be represented mathematically; it may then be possible to derive analytical expressions to describe behavioural sequences. Computer simulation allows the investigation of more complex cases, by simulating both the animal and the environment, and so allowing the external consequences of the model animal's behaviour to be fed back to the model animal. (In this type of work, the simulated animal-like agent is often called an animat). When carefully done, this is an extremely valuable technique. However, it depends on the use of adequate models both of the environment, and of the way in which the environment determines the input to the animal's control system, and such models are not always available.

This technique of simulating a mobile agent in an environment is also used widely within the engineering discipline of robotics. In fact, a look through the papers in many robotic journals and conferences will reveal that studies of simulated mobile robots are often more common than those dealing with real mobile robots, and that it is quite usual to refer to simulated robots as robots, without qualification. There are two underlying reasons for the popularity of these simulation studies: they can be quick, and they can be cheap. Constructing a new robot is normally both slow and expensive; if tests show that some aspect of the robot needs to be changed, implementing the changes may require further substantial amounts of time and cost. Robot simulations allow much of the design testing and modification to take place before construction begins; they can also reveal that an idea is unfeasible, and so avoid a wasted construction phase altogether. Once again, the use of simulation depends on the availability of adequate models of the robot and of the environment.

For ethology, there remains one type of technology-based approach that holds out the prospect of providing both an attractive analogy and a solution to some of the difficulties of computer simulation: the use of real mobile robots. Technical progress over the last fifty years has now brought us to the point where it is possible to build robots that can model many aspects of animal structure and function in sufficient detail to be useful to ethology, as can be seen from the case studies in this volume. This chapter provides a background for the case studies by reviewing some landmarks in the history of animal-like robots. It also explores some of the ways in which robotics and engineering may be subject to biases and constraints that may limit the analogical or investigative use of robots within biology.

Robots and animals: similarities and differences

Mobile robots have now existed for some fifty years. During that time, most robots were developed within the technical and conceptual horizons of contemporary engineering, computer science, and artificial intelligence, and only a small number, for a variety of reasons, were specifically designed to resemble animals in one way or another. Now it happens to be the case that animals and mobile robots, whether animal-like or conventional, necessarily share so many features that, in many ways, all mobile robots resemble animals to some extent. On the other hand, all robots, whether animal-like or conventional, have things in common that set them apart from animals. Clearly, any assessment of the merits and deficiencies of a particular animal-like robot should be made in the context of these generic similarities and differences, which are outlined briefly in this section.

Similarities

Both animals and robots have bodies with movable parts (effectors), which they use to move around within an environment, and which they also use to act upon the environment. Both make use of sensors – specialized structural elements that are affected by specific local qualities within the environment. And both have other structural elements – the components of the control system – that can be affected by the sensors and by internal factors, and that affect or control the movable body parts.

Since sensors and effectors involve physical elements interacting with the physical environment, it is not surprising that there are further parallels. Indeed, animal structures associated with particular problem domains are often used as a source of inspiration by engineers (French 1988). Many sensors in both animals and robots respond to the same sensory modality – heat, electromagnetic radiation of certain wavelengths, sound and vibration, salinity, gravity, pressure, and so on. Because there are often good and simple mechanisms for dealing with each modality, similar designs are often seen in biological and engineered systems. For example, the lens and iris of the eye have almost exact structural and functional parallels in many cameras. In the case of bodies and effectors, there are often obvious differences between the engineering and biological realms: for example, animals do not use wheels, and there are almost no analogues of muscle in robots. However, there is significant overlap in the areas of mechanical linkages and mechanical structures, with the modern use of composite materials echoing many aspects of skeletal bone structures.

It is also easy to see some common factors in the low-level elements of engineering control systems and in those of biological control systems. For example, both face the problem of sending signals over long distances without significant degradation due to noise. In modern communications systems, this is achieved by sending binary signals via a number of successive stages, rather than travelling the whole distance in a single hop; a repeater at each stage regenerates the signal locally to prevent attenuation, and consequent loss of information as the signal to noise ratio rises. In the myelinated axons of vertebrate neurones, the information in the signal is again encoded in a binary form, and is again transmitted not from end to end directly, but by stages between successive nodes of Ranvier, at each of which the signal is again locally regenerated.

It is also possible to see common principles at work at a higher level – for example, in the modes of locomotion of animals and robots (and of some remotely controlled devices that are not independent enough of the operator to qualify as robots). There are only so many different types of terrain or medium, and there are only so many ways in which coherent movement is possible within each type, so a degree of convergence is not surprising. Any animal or robot that uses legs for locomotion must move those legs in one of a relatively small set of possible gaits if it is to move in a stable and efficient way. Insects have six legs, and almost all use the alternating tripod gait. Most legged robots also have six legs, and also use the alternating tripod gait; this is not because there is any intrinsic advantage in copying insects, but because this combination of legs and gait is stable and economical – it is a fundamental property of the domain of legged locomotion, rather than being fundamentally animal-like.

Some such principles are only discovered when a mode of animal locomotion is imitated by engineers for some reason. For example, both marine engineers and biologists have long been intrigued by Gray's paradox, which observes that many fish can swim faster than conventional theory predicts they should, in that calculations show that the muscles of the fish are not powerful enough to drive a structure with the measured drag of the fish at the measured speed of the fish. Recent experiments with RoboTuna, a model tuna which copies the body shape and swimming movements of a real tuna, have shown that the measured drag while swimming is reduced to about 40% of that while static; the corresponding figure for a real tuna is estimated at about 15%. The combination of shape and undulatory motion appear to give rise to some presently ill-understood but beneficial hydrodynamic effects; further investigation and exploitation of this phenomenon is now being undertaken both for scientific reasons, and for the propulsion of marine robots and boats. But this is once again clearly a property of a domain, and is animal-like only in a weak sense.

Differences

A major cause of differences between animals and robots is that the components and systems of animals – the sensors, effectors, and control systems – are evolved and grown; in contrast, conventionally engineered components and systems are designed and built. It will be convenient to examine these factors separately before summarizing their significance.

Evolved systems

Evolved systems develop by a process of progressive and often non-linear variation of previous evolved systems, all of which were good enough to ensure that their owners bred successfully. (For an extended examination of the nature and consequences of the evolutionary process, see Dennett [1995].) This variation often takes the form of adding components to an existing system, so that the functions performed by a given structure before variation are performed by what is essentially the same structure afterwards; new functions are carried out by the new set of components, working in the context of the older components and functions. A structurally separated functional unit like this can be regarded as a module. Similar brain structures performing similar functions can be seen throughout the vertebrates, regardless of the different additional structures and functions present in different species. However, evolution is also parsimonious: additional structures impose additional costs in specification, growth, space, and energy supply, and the modification of existing structures to perform different or additional functions is also common. This can be seen in many physical structures – perhaps the best known example is the way in which parts of the jaws of earlier vertebrates have evolved into the 'chain of tiny bones' that conducts sound from the eardrum to the cochlea in mammals. And although evolved structures are often elegant and clearly intelligible solutions to particular problems, they are not constrained to be so; all that is required is that a variation works better than the competition. Working well enough, often enough, and being cheap enough are the only necessities; elegance and intelligibility are happy accidents, perhaps encouraged by severe pressure for economy and hence simplicity.

Evolution does not have any capacity for lookahead, or for deferring gains; what works in the current evolutionary situation will win, regardless of what opportunities for further development are closed off by this 'choice'. Perhaps the best known consequence of this is in the vertebrate eye. The sensory cells in which light is converted into nerve signals are positioned at the back of the retina; the nerves carrying information from the eye to the brain originate from a layer in front of the sensory cells. The information-carrying nerves are gathered together and leave the eye in a single bundle, the optic nerve; at the point at which they leave the eye, there is no place for the sensory cells, and so this location – the blind spot – is insensitive to light. In the octopus and many other invertebrates, on the other hand, the sensory cells are in front of the layer made up of the information-carrying cells, and so the sheet of sensory cells is continuous across the exit point of the optic nerve, and the octopus has no blind spot. The roots of this difference lie far back in evolution, when the branch leading to the vertebrates happened to take one of the many approaches leading to useful light-sensitive organs, and that leading to the octopus took another. (For an excellent discussion of the evolution of eyes, see Dawkins [1996].)

A consequence of this lack of lookahead, and of the way in which the variations giving rise to solutions are essentially random, is that evolution is often best thought of as proceeding by a mixture of hacking and tinkering (or *bricolage*), and the results often bear the hallmarks of the process. The *bricolage* is not applied to a single component in isolation, but to all components – sensors, effectors, and control systems – at the same time. Since the behaviour of any system – animal or robot – is a complex emergent function of these three elements and their interaction with the environment, a beneficial change in behaviour may not be attributable or reducible to a particular change in any one of these elements. It is no surprise that the application of artificial evolution to the development of robot sensors and control systems also produces spectacularly untidy but effective results, which often resist analysis.

Grown systems

A system that has to be grown is both constraining and constrained. Biological growth processes are similar to evolutionary processes, in that development necessarily builds on what has previously developed, and so structures are often layered, although spatial distortion can make this layering difficult to discern. However, the methods often used to direct growth – the setting up of gradients by diffusion or other processes – are very good at making approximate connections, but are poor at detail. The fact that growth often continues while the system is functioning is a further complication. In different animals of a given species, growth as a whole or relative to other parts may proceed at different rates, yielding systems with components of different sizes and different relative proportions. This applies to the nervous system as well as the sensors and effectors; animals such as some crustaceans, which always have the same numbers of neurones in the same places, and with the same connections, are rare.

The upshot of all this is that, to be functionally accurate and precise over the lifetime of the animal, grown systems must usually be adaptable: structures that are required to have some exact structural or functional relationship with other structures, or to represent some variable or parameter with a high degree of accuracy, must be capable of being fine-tuned after the initial growth has taken place. Sometimes this can be done by exploiting the growth process itself – for example, by arranging that only components in the 'correct' relationship to other components continue to grow, or even to survive. A phase of apparently excessive initial growth, followed by a stage of pruning, is found in many nervous systems, including the human brain. The 'correct' relationship may be defined in terms of receiving a certain sort of stimulation or combination of stimulations from the other components. If the available stimulation is determined by environmental factors, then the structures formed, and hence the functional characteristics, may be shaped by those factors. The surprising extent to which such processes may influence the development of the human brain is set out in the theory of neural constructivism (Quartz and Sejnowski 1997); some consequences of the dynamic adjustments made when body parts are removed or congenitally absent are described in Ramachandran and Blakeslee (1998).

Modularity in evolved and grown systems

As noted above, a typical feature of naturally evolved and grown systems is modularity – the appearance of identifiably separate structural and/or functional components. Structural modularity is easy to recognize. It occurs in two main forms: as units, often similar to one another, making up another structure (for example, the ribs making up the rib cage); and as one or more isolated units with some identifiable function (for example, the heart, or the kidneys). Functional modularity without any clearly corresponding structural aspect is a more difficult notion, and is often referred to in the context of behavioural abilities or brain function (Pinker 1997), where the nature of the underlying structure is unknown. It is often possible to decompose an animal's behavioural abilities into a number of functions, and to establish by experiment the determinants of performance levels within each such function. Unfortunately, it is sometimes forgotten that such an analysis by itself does not imply that there are functional modules within the animal that correspond to these functions; still less does it imply that there are structural elements corresponding to these functions. However, there are undeniably many instances in which the search for a structural unit corresponding to a hypothetical functional unit has been successful.

A particularly common type of modularity occurs when a structure consists of a number of similar substructures, each of which consists of further substructures, and so on. This is an instance of hierarchy, a very general biological principle which applies not only to structural factors but also to the functional aspects of many systems (Dawkins 1974).

In both structural and functional modularity, the observed separation of a unit implies a degree of independence, in that it may be possible to modify a module without the process of modification directly affecting another module. This may allow evolution to act on one aspect of structure or function while leaving others substantially unaffected. As the underlying genetic basis of evolution is itself modular, it is perhaps not surprising to see modular features in evolved systems. There are also grounds for expecting modularity in grown systems, as many processes in morphogenesis naturally give rise to repeated, layered, branched, or enclosed structures.

Designed systems

Conventionally engineered systems are designed, usually subject to economic considerations. The extent to which these factors impose constraints on what can be produced is not generally appreciated by those outside the profession. The design process has developed to the point where it is now quite stereotyped; to some extent this has been enforced by the insistence of governments and other large customers that the performance of products must be assured by the knowledge that they have been designed according to what is held to be good practice. The first step is to prepare a specification of the features and performance attributes required of the system. This stage often involves analysis — the identification of explicitly defined variables and functions thought to be fundamental to the system definition. If the system is complex, it may then be broken down further into functional and structural subsystems, each of which will require a specification; the interfaces between these subsystems must also be explicitly defined. The nature and number of the subsystems may be determined by the existence of specialist teams of engineers familiar with subsystems of particular types – power supplies, signal processing, motors, and so on. On the other hand, physical or other constraints within the system itself may dictate that a particular subsystem needs to be broken down further. Large subsystems may be broken down into smaller subsystems, until a level is reached at which detailed design can take place. Note that this produces a system that will almost certainly have some hierarchical characteristics, both in structure and function. In some international standards, there is a requirement that critical stages of design should be supported by calculations carried out by two entirely different methods; this implicitly restricts the design approach to one in which design variables must be explicit and quantifiable. Throughout this process, the two variables of cost (both of design and manufacture) and elapsed time may influence any decision.

Where possible, the components and subsystems used in the design will be stock items, already manufactured in quantity to known specifications. The variations to be expected in these components will be known; the design will normally be guaranteed to meet specification even if the worst-case combination of component variation is encountered. It is often appropriate to use a component that is far better than the design requires, if it is cheap and available. For example, there is really no necessity for something as simple as a toaster to be controlled by something as complex as a microprocessor, but since cheap microprocessors and the infrastructure to use them already exist, it is often impossible to justify any other approach.

The subsystems will be assembled and individually tested, and will then be assembled into the final system, which will be tested again. The design process specifies the precision of assembly required to achieve a defined performance. Any subsystem failing a test may be repaired, or replaced by another working example.

The design practice just described expresses the modern philosophy of mass production and quality systems: both systems and parts are made to defined standards, and parts that must work closely together are individually made to corresponding specifications that ensure that they will fit within defined limits. This method of design and manufacture replaced the craft system, in which parts were individually made to fit one another by a master craftsman, as a result of which no two systems ever had a defined structure or performance, and parts were not generally interchangeable.

The development of software follows much the same path as that described above. Large systems are decomposed into subsystems of manageable size by a process of top-down analysis, existing pieces of software are reused where appropriate, and systematic methods of design ensure consistency and reliability.

Built systems

Engineered systems and components are built rather than grown. Modern assembly and manufacturing technologies have been developed along the same lines as the design process, with the repeatable achievement of defined standards, and the measurement and control of sources of variability, forming the basis of most processes. This means that impressive degrees of precision in manufacture and assembly can be achieved, and so systems in which every one of tens of millions of components and connections must be correct, as is the case with integrated circuits, can be successfully constructed. Just as evolution can only produce systems that can be successfully grown, the engineering process can only produce systems that can be successfully built. In the present state of technical development, it is difficult to see that manufacture and assembly are acting as a significant constraint on design.

The control of sources of variation means that there is now little requirement for any processes of adaptation. A few decades ago, engines had to be run at less than full load for a period (running in) to allow deliberately ill-fitting bearing surfaces to wear slightly to improve the fit and the load-bearing capacity. This is no longer necessary, as the required precision can now be achieved in manufacture. However, some low-level capacity for adaptation is often provided for mechanisms linked to components that are designed to wear rapidly and then to be replaced; examples are automatic brake adjustment systems.

Engineered systems also require maintenance and repair. In the past, this was a significant constraint on design, manufacturing, and assembly, as systems had to allow partial disassembly and reassembly for the repair or replacement of individual components that were worn or defective. The modern practice is to replace subassemblies rather than components, and so the constraints are fewer and less onerous. In some cases, continuing cost reductions make it economic to do away with the idea of repair altogether, with entire systems being regarded as disposable, in that, if they fail, they are simply replaced with a whole new system. This effectively removes the repair constraint from the design, manufacturing, and assembly processes.

Modularity in designed and built systems

Modularity in designed and built systems is pervasive. It arises at many stages: in the analytical process; in the technical management processes of allocating design tasks to different teams; in the economic advantage of reusing existing subassemblies; and in the requirements for assembly and maintenance. However, it should be recognized that these are all factors to do with the organization of the engineering profession; it is not necessary for an engineered device to be modular, but it is quite unusual for it not to be so.

Significance of similarities and differences for animal-like robots

There are a number of lessons that can be drawn from the factors considered above. The similarities between animals and robots imply that, in some cases, a robot may appear to be a good model of an animal simply because they are both solving the same problems in the same way using similar components. This essentially gives us knowledge of the domain, and may do little more than confirm that evolution and engineering can both converge on the same good solutions. The differences tell us that in some cases it may be difficult or even impossible for a robot to imitate an animal because the animal uses functions or components that are alien to engineering practice.

The existence of functional and structural modularity in both natural and engineered systems seems to offer a degree of compatibility, but it may also be deeply deceptive. If an adequate functional analysis is performed on a certain aspect of animal behaviour, then it may well be possible to synthesize a robotic system that implements this analysis and appears to mimic the animal's behaviour. However, unless additional evidence is produced to show that there are correspondences between the functional modules in both animal and robot, the success of the robot cannot be taken to imply any understanding of how the animal produces the behaviour. This realization is particularly important because many robots make a strong impact on the observer, and one must guard against being convinced by the quality of the behaviour alone. A particularly interesting situation occurs when a robot is developed that can adequately mimic a behaviour while having no internal modules corresponding to the functional decomposition of the behaviour proposed by biologists; an example of this can be seen in Case Study 2 in this volume.

We can now examine the evolution of animal-like robots over the last fifty years while keeping in mind the concepts developed above. Progress has been decidedly non-linear. As often happens in science and engineering, a single early effort – that of Grey Walter in the early 1950s – sketched out many of the enduring issues in the field, and yielded insights that were subsequently lost, only to be rediscovered decades later. We shall begin with a review of his work.

The contribution of Grey Walter

Grey Walter was what would now be called a neuroscientist. He had trained as a physiologist, initially studying conditional reflexes, and then becoming involved in the new technique of electroencephalography, to which he made many original contributions. He was particularly adept at developing new equipment; although he was not himself an engineer, he had an excellent understanding of the electronic technologies of the day. He wished to produce a 'scientific imitation of life', which would be 'concerned with performance and behaviour', and which would enable him 'to learn about life by imitation as well as observation of living things' (Walter 1953).

Recognizing that 'the brain's elaborate performance' depended on a high level of intrinsic complexity, but that the vast number of brain cells was an obstacle to modelling the brain at the cellular level, he cast around for another strategy. He found it by realizing that the number of different combinations of activity ('modes of existence') obtainable from a relatively small but richly interconnected set of units offered a simple means of obtaining complexity. By allowing each unit a single directional connection to each other unit, and by assuming that a given unit could 'drive' another unit via such a connection, he calculated that 'With six units there would be enough modes to provide a new experience every tenth of a second throughout a long lifetime'. This led him to speculate that 'the elaboration of cerebral functions may possibly derive not so much from the number of its units, as from *the richness of their interconnections*', and to propose that 'some elementary experience of the actual working of two or three brain units might be gained by constructing a working model in those very limited but attainable proportions'. He continued: 'But if the performance of a model is to be demonstrably a fair imitation of cerebral activity, the conditions of stimulation and behaviour must equally be comparable with the brain. Not in looks, but in action, the model must resemble an animal'. The characteristics of behaviour he thought necessary included 'exploration, curiosity, free-will in the sense of unpredictability, goal-seeking, self-regulation, avoidance of dilemmas, foresight, memory, learning, forgetting, association of ideas, form recognition, and the elements of social accommodation. Such is life'.

In 1949, he built two 'model animals', the 'tortoises' or 'turtles' Elmer and Elsie, very similar in all crucial respects but differing in the detailed implementation of various features. Although they worked, they were too unreliable and fragile to cope with repeated demonstrations, and in 1951 six more were built, to a similar but much better engineered design (Figure 2.1). He claimed that the basic design contained the equivalent of only two 'brain cells', in the form of two thermionic valves (vacuum tubes); however, it also contained two adjustable single-pole changeover relays – electrically controlled devices capable of switching a connection to either of two contacts as a function of the relay current and the positions of the adjustable contacts – which arguably could also correspond to brain cells. Two sensors were provided: a directional light-sensitive device producing a graded output, and an omnidirectional touch sensor operating as a switch. The three-wheeled chassis was laid out like a child's tricycle, and consisted of a single front wheel, steered and driven by two independent motors, and two fixed rear wheels. When the steering motor was on, it rotated the front wheel about its vertical axis, and always in the same direction, giving the system some built-in asymmetry. The light sensor was mounted on top of the structure that steered the front wheel, and was aligned with the wheel so that it always pointed in the same direction as the wheel. As the steering motor rotated the wheel about its vertical axis, it caused the light sensor to scan round the environment like the light in a lighthouse. The whole robot was encased in a shell which was coupled to the touch switch so that sufficient pressure from any direction would close the switch. A lamp on the front of the tortoise was lit whenever the steering motor was on; the lamp could

Figure 2.1. *A Grey Walter tortoise from 1951.* This example, now in the Science Museum, has a clear Perspex shell allowing the main features of the mechanism to be seen. The photo-electric 'eye', protruding through the shell on the right, is looking through its aluminium shroud straight at the camera; note its alignment with the steered and driven front wheel. One of the two thermionic valves is inside a second shroud projecting just in front of the rear wheel in the foreground. The H-shaped arrangement supporting the shell at the top is mounted on a switch operated when the shell is deflected. The curved lever on the left allows the tortoise to be switched on or off without having to remove the shell; it may also be operated by a floor-mounted projection inside the hutch to switch the tortoise off automatically as it moves into the hutch.

not affect the photosensor other than via a mirror, but could affect the photosensors of other tortoises.

Grey Walter described the behaviour of the 'tortoises' at the time, but the terms he used do not relate well to modern ideas. However, in notes written in around 1961 but not published until very much later (Holland 1996) he describes and analyses their structure and behaviour in terms that are strikingly modern, and that serve as a useful introduction to modern thought. The tortoise had four mutually exclusive basic 'behaviour patterns':

Pattern E: exploration. When the light falling on the photosensor is below some threshold (the lower threshold) and the shell is not in contact with an obstacle, the drive motor is turned on at half speed and the steering motor turned on at full speed. The robot moves through the environment in a cycloidal path – a kind of looping sideways trajectory.

Pattern P: positive tropism. When the light falling on the photosensor is above the lower threshold, but is below another higher threshold (the upper threshold), and the shell is not in contact with an obstacle, the drive motor is turned on at full speed but the steering motor is switched off. The robot moves quickly along an arc defined by the angle at which the front wheel happened to be when the steering motor was stopped. If by chance the wheel was aligned with the axis of the robot, the robot will begin to move directly towards the light (assuming the light itself does not move or is not obscured). In all other circumstances, the curved trajectory will change the orientation of the photosensor, causing the photosensor stimulation to drop below the lower threshold, and terminating behaviour pattern P.

Pattern N: negative tropism. When the light intensity at the photosensor exceeds the upper threshold, the drive motor is turned on at full speed and the steering motor at half speed. This ensures that the robot steers away from any light source bright enough to produce this behaviour pattern; this will of course terminate the behaviour pattern as soon as the orientation of the photosensor is no longer in line with the direction of the light source.

Pattern O: obstacle avoidance. If the shell is sufficiently displaced by contact with an object, the control circuit goes into oscillation and alternates between the motor outputs of behaviour patterns E and N at about twice per second. This produces successive bouts of what Grey Walter called 'push-hard-steer-gently' and 'steer-hard-push-gently', which is very effective at both pushing light movable obstacles out of the way, and gradually changing the direction in which the robot is pushing until it is able to move away freely. In either case, once the robot is no longer in contact with the object, the oscillation ceases and the behaviour pattern terminates. While O is operative, the photosensor input has little or no effect, regardless of the light intensity.

These behaviour patterns were produced by the action of the basic circuit in conjunction with specific component values, and by the fine adjustment of the relays. Observation of the surviving original tortoise, of replica tortoises, and of original film of the tortoises, shows that it is easy to identify which basic behaviour pattern is active at any time. However, Grey Walter's real achievement was in showing that these basic behaviour patterns could be sequenced and combined to produce what appeared to be coherent higher level behaviours. The sequencing and combination emerge from the relationships between the basic behaviour patterns imposed by the circuit generating them, and by the interaction of the basic behaviour patterns with the sensory consequences of the robot's movements through the environment.

The nature of the relationships between the behaviour patterns is quite simple. The patterns fall into two groups. The first consists of E, P, and N. One of these operates whenever the shell is not displaced; which one is active is determined by whether the photosensor illumination is below the lower threshold, between the lower and upper thresholds, or above the upper threshold. The second group consists of O; it operates when the shell is displaced, regardless of the photosensor input. Grey Walter regarded O as being an example of a prepotent reflex, one that, when stimulated, prevents the expression of other reflexes currently receiving stimulation sufficient to trigger them. Suppose a tortoise is heading towards a light source using pattern P, and encounters an immovable obstacle that does not obscure the light. If P were prepotent over O, the tortoise would become stuck, and would continue attempting to move towards the light until its batteries ran down. However, since O is prepotent over P, the tortoise will 'butt and turn' until it is no longer in contact with the obstacle; the nature of the movement is such that, when the behaviour terminates, the tortoise is most unlikely to be in exactly the same position and orientation as on its first approach, and so will approach the light on a different course. If this course again brings it into contact with the obstacle, the cycle will repeat; eventually the tortoise will find itself with a clear run towards the light. If the obstacle is too light to trigger the shell switch, it will be bulldozed aside by P; if it is heavy enough to trigger the switch, but light enough to be moved by the tortoise, it will be moved by O; if it is either of these, but becomes jammed against some other object while being pushed away, the tortoise will respond as with an immovable object. Every possible combination of obstacle type, even a change from movable to immovable or *vice versa*, is responded to with a sequence of behaviour patterns that allows the tortoise to continue towards the light.

It is also easy to see how basic behaviour patterns can be appropriately sequenced by their interaction with the sensory consequences of the robot's movements through the environment. Although P, when triggered, locks the tortoise onto a course that initially moves it towards a moderate light, the curved course means that it will reach a point at which the change in orientation of the sensor leads to the photosensor output falling below the lower threshold. This will lead to the production of E, which will make the photosensor scan round until it points at the light and P is again engaged.

There are two quite different ways in which this can happen. Assume that the sensor rotates anti-clockwise. If the axis of the tortoise is pointing to the left of the light at the instant when P is engaged, the axis of the photosensor must point to the right of the axis of the tortoise, and the machine will move along an arc curving to the right. This will move the axis of the photosensor to the right of the light, causing E to appear; E will rotate the axis of the sensor to the left, realigning it with the light source, and triggering P again. This alternation can happen so rapidly that the tortoise appears to be moving smoothly towards the light source along a path slightly curved to the right. Let us call this type of movement M1. It is sometimes possible for the system to become locked on the light source, moving towards it in a straight line, but more usually it strays to the right of the light, when a new type of trajectory appears.

If the axis of the tortoise is pointing to the right of the light when P is triggered, the tortoise will move on a leftward curving arc; as the axis of the

28 *Artificial ethology*

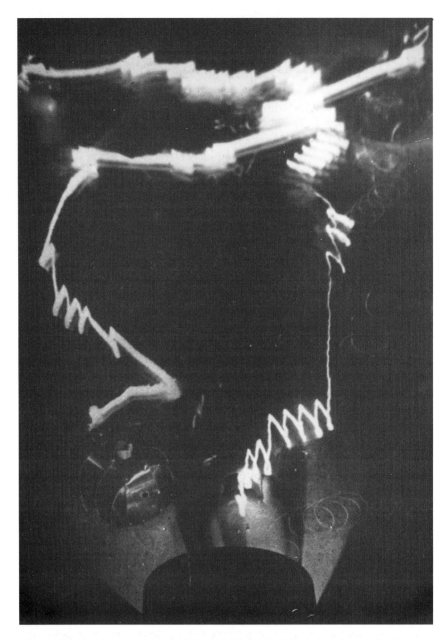

Figure 2.2. *Time lapse photograph of Elmer and Elsie.* For this photograph, which dates from 1949, Elmer and Elsie were fitted with shell-mounted candles to trace out their movements.. Elsie starts at the top left, crosses to the right, and moves towards the light in the hutch (bottom centre); Elmer starts at the top right and crosses to the left.. Note the alternations of long straight runs (M1) with zig-zag sections (M2) in the traces of both robots; note also the interaction between the robots (top centre) and the slight differences in character between the traces of Elmer and Elsie. The faint circular traces are from the pilot lights.

In 1949, records of the tortoises' behaviour were made by fixing candles or lights to the shells, and taking time-lapse photographs. One such photograph, Figure 2.2, clearly shows the alternation between M1 and M2. Each moves the tortoise towards the light; the less efficient and slower M2 also scans the entire environment at each iteration, allowing the tortoise to respond to any other lights in the environment, rather than remaining fixated on one alone.

We thus see that, in an environment containing a moderately bright light, the interaction of the basic behaviours with the sensory consequences of the tortoise's movements produces trajectories consisting of two clearly identifiable movement patterns, each made up of the alternation of two basic behaviour patterns. It would be perfectly reasonable to regard the approach to the light as a single high-level behaviour, consisting of alternations of the two lower-level behaviours M1 and M2, each of which consists of alternations of the basic behaviours E and P. However, only the basic behaviours are built into the system; the others are what is now called 'emergent', appearing in particular environmental contexts as a result of the interaction between the basic behaviours, the environment, and the effects of the basic behaviours on the environment and on the robot's relationship to the environment. For a modern discussion of this type of emergence, see Steels (1994).

The tortoise demonstrated several other behaviours in different circumstances. In the single light source situation, a light that at a distance produces stimulation between the lower and upper thresholds, showing the behaviours discussed above, may produce stimulation above the upper threshold when the tortoise is close enough. This will trigger alternating bouts of behaviour patterns N and E until the tortoise has moved far enough away for alternating bouts of P and E to occur, when it will approach again. The asymmetry of the scanning mechanism means that the tortoise will move around a suitable light source in an irregular circle; such trajectories contain smooth sections and jagged sections, just like the approach trajectories (Figure 2.3). Grey Walter thought of this behaviour as 'seeking optima', pointing out that it was effectively a system with negative feedback, each behaviour correcting deviation from some equilibrium position. With two light sources spaced at a suitable distance, it will approach and circle first one, and then the other; as can be seen from Figure 2.4, it escapes from the one it is circling when its alignment enables it to make a long enough run without making a full 360° scan, using a type M1 movement. Grey Walter thought this situation important, as it represented the classical dilemma of Buridan's ass, which starved to death when equidistant from two bales of hay.

These demonstrations of emergent behaviours deliver two messages. For biologists, they imply that the observation of an identifiable behaviour does not mean that there is any structure or module responsible for the production of that behaviour. For engineers, they show that useful and apparently flexible behaviours do not have to be explicitly designed. The drawbacks of emergent behaviours, which make them unsuitable for use in conventionally engineered

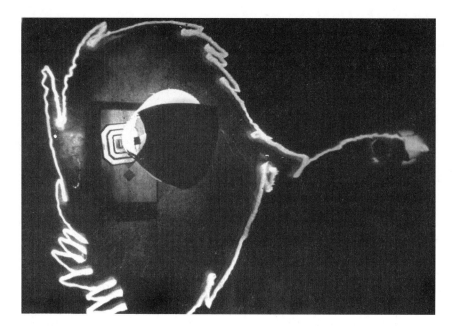

Figure 2.3. *Elsie approaching and circling a light.* This photograph is taken from above; the lamp is shielded from the camera. Elsie is released on the right, and approaches the light by alternating behaviour patterns E and P. When close to the light, N is triggered whenever the photosensor is aligned with the lamp; Elsie then produces a sequence involving behaviour patterns N, P, and E, which produces the effect of circling the light. Note the presence of relatively straight sections (M1) and zig-zag sections (M2) in the circling behaviour.

systems, are almost never mentioned: there is no design procedure; proof that the behaviours will always work appropriately may not be available; and a fault in a basic behaviour will affect all the emergent behaviours that it supports.

There is a rather different sense of emergence which also appeared in Grey Walter's work. Rather than producing emergent behaviours intentionally from known basic behaviours, it is possible for them to arise unexpectedly because of some unintended change in the robot itself, or some unanticipated environmental effect. In engineering, the effects produced by such factors are regarded as faults or bugs, and steps are taken both to prevent their appearance, and to ensure that, if they do appear, the system is changed to prevent their appearance in future. However, if they appear in the context of a robotic system that is already exploiting emergence, and if the unanticipated behaviours that emerge are potentially useful, it makes sense to attempt to capitalize on them.

A by-product of the tortoise circuitry was that, as time went on and the battery voltage fell, the upper threshold would rise. This caused the tortoise to approach much closer to strong lights. Grey Walter noticed this, and seized upon it to produce the first example of a robot that was able to recharge its batteries as they became exhausted. He built a hutch containing a battery-charging circuit,

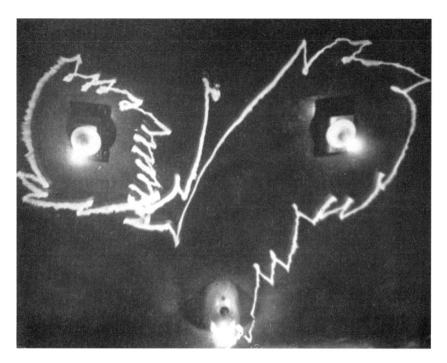

Figure 2.4. *Elsie solves the dilemma of Buridan's ass.* Elsie is released at the top, midway between two lamps. The robot first approaches and circles the lamp on the left, then produces a long straight movement (type M1) which brings it within the influence of the lamp on the right, which it circles once before moving off to the bottom of the picture.

and arranged contacts on the tortoises and on the hutch so that a tortoise entering the hutch would be connected to the battery charger. He positioned a bright light inside the hutch so that a tortoise with a fresh battery would be repelled by it when about to enter the hutch (behaviour pattern N) but one with a discharged battery would be attracted (behaviour pattern P). The high current flowing in the charging circuit was arranged so that it would switch off the power to the tortoise's sensory and motor circuitry. Once the battery was sufficiently charged, the charging current would fall, and power would be restored to the tortoise's control system. The tortoise would be repelled by the light and would begin to move, breaking the contact with the charging circuit, and leaving the hutch, to return when the battery was again exhausted. Although far from reliable, this arrangement produced a robot that could achieve a degree of autonomy within its environment, at least in principle.

Further instructive examples of emergence were provided by the tortoises' headlamps – the small lamps that had originally been added to indicate when the steering motor was taking current (during behaviours E and O). A tortoise producing E in front of a mirror would be switched to P by the reflection of its own lamp, and would begin to approach the mirror; however, since the light is not lit during P, it would switch back to E, which would relight the lamp, inducing the tortoise to switch to P and approach again, and so on. The outcome of this looked like an additional behaviour; Grey Walter reported that the tortoise '...flickers and jigs at its reflection in a manner so specific that were it an animal a biologist would be justified in attributing to it a capacity for self-recognition'. Two tortoises facing each other '...would be affected in a similar but distinctive manner', a further behaviour which Grey Walter was happy to call mutual recognition. (Of course, modern biologists would not feel justified in using these terms.)

A final observation arose when a tortoise was put in an environment containing a lamp, and several low and easily moved obstacles. As it repeatedly moved towards and away from the lamp, and explored the space, it would move the obstacles out of the way, 'and sometimes seems to arrange them neatly against the wall'. The behaviour of the tortoise changed the environment in such a way that its subsequent behaviour was affected: it could approach the lamp without encountering obstructions. This is the first report of a weak form of what is nowadays called stigmergy. (See Case Study 3 for some recent work on stigmergy in robots.)

Artificial intelligence and control engineering

In the thirty-five years after Grey Walter's tortoises, considerable progress was made in biology and ethology in understanding what problems were faced by animals in controlling their behaviour, and how such control was achieved. However, there was very little activity or progress in the area of animal-like mobile robots, apart from some extensions of Grey Walter's ideas, and a brief flurry of machines modelling the learning of mazes by rats. The main reasons for this lack of progress were the increased use of computers, the development of computer simulation techniques, and the associated rise of artificial intelligence. Computers were the tool of choice for any type of intensive computation, but they were large, heavy, expensive, and hence immobile. Some civilian research robots were made; they were tethered to the host computer by an umbilical cable delivering sensor data to the computer, and motor commands to the robot; none were animal-like in any meaningful way.

One way out of the impasse was to use the computer itself to simulate a robot and an environment, and to control the simulated robot. As we noted above, where the environment, the effect of the environment on the robot, and the effect of the robot on the environment are adequately characterized, this is an excellent technique. However, the environments that allow such characterization are all essentially empty: space, air, and water. Animal-like robots are essentially terrestrial, and all simulations of terrestrial worlds during this period, and up to the present day, were very idealized.

The main reason why few animal-like robots were developed during this period is probably that the dominant paradigm was artificial intelligence (AI), which had set out to mimic human thought processes and human intelligence. In the context of robots, AI held that intelligent behaviour would result from the sensing of the environment, the construction or updating of a symbolic internal representation of the environment (the world model), the application of symbolic reasoning to the problem of achieving the task in the context of the environment, and the execution of whatever behaviour was indicated – the so-called sense-model-plan-act cycle. Unfortunately, every aspect of this vision turned out to be problematical: the world could not be adequately rendered into a symbolic form; reasoning across the internal representation of the world did not produce appropriate results; and symbolic commands for action could not be satisfactorily implemented. Although most researchers in AI believe that these difficult problems will one day be solved, there is a body of opinion that holds that they are intrinsically insoluble within the AI paradigm; a typical representative of this point of view is Dreyfus (1992).

In contrast to the travails of AI, control system engineers producing conventional robots made great strides in this period. Many aircraft, civil and military, were effectively transformed into robots, as were most spacecraft and many ships. Industrial robots, including some mobile robots (AGVs – automated guided vehicles), were outstandingly successful at a range of stereotyped tasks. There are parallels between some of these systems and certain aspects of animal behaviour. For example, basic industrial robots carry out a predetermined sequence of actions without sensing what effects the actions have in the world, or how the world has changed. This works well enough in the highly engineered and unvarying environment of the factory. It is also a strategy used by insects such as the digger wasp; parts of the behavioural repertoire, once triggered, execute inflexibly regardless of whether the actions are achieving appropriate results (Baerends 1941). The control systems used in aircraft and guided missiles often involve simple negative feedback, as seen in many animal systems.

Rodney Brooks and behaviour-based robotics

It is usually accepted that the real advances in animal-like robots came after 1986, with the introduction by Brooks of the subsumption architecture (Brooks 1986), and the development of behaviour-based robots. Disappointed with the contribution to robotics of conventional artificial intelligence, Brooks proposed a radically new approach to building 'intelligent systems', inspired in part by biological considerations. Rather than decomposing an intelligent system along functional lines – a perceptual component, a central representation, a reasoning component, an action component, and so on – he adopted a strategy of dividing it into a number of behaviour-producing subsystems, each independently connecting sensing to action to achieve some particular behavioural competence. Typical subsystem behaviours were Avoid (which avoided obstacles), Wander (which kept the robot moving, changing direction every so often), and Explore (which looked for long clear paths and directed the robot towards them). The amount of internal state within subsystems was kept low, as internal state was held to be equivalent to internal representation, which Brooks was determined to avoid. The subsystems were essentially in parallel with one another, and could be thought of as forming a layered structure. The system was developed incrementally, with the lowest layer being designed, built, and debugged before the next layer was added; the idea was that a layer, once working reliably, should be 'frozen', thus ensuring that the behavioural competence that it provided was always available, even if higher levels failed.

Since each layer was developed in the context of the previously built layers beneath it, the correct functioning of a layer often relied on the existence and integrity of lower layers, but was always independent of the existence of higher layers. The influence that a layer could exert on lower layers could take two forms: suppression, which injected a signal into the input side of a lower layer component in place of some existing signal; and inhibition, which inhibited the output of a lower layer component, but did not replace it with a new signal. This interference with the lower layer was described as subsumption, as the lower layer was then in effect treated as a part of the higher layer. However, the term 'subsumption architecture' was often used by others to describe a much simpler arrangement, in which layers were given some fixed priority, and the activated layer with the highest priority inhibited the outputs of all lower levels. (In fact, the most complex robot using this approach was designed within Brooks's lab, by Connell (1990), who was careful both to differentiate this architecture from the original formulation of subsumption, and to justify it on engineering grounds.) This simpler approach abandoned the incremental origin of layering, and also the idea that a layer could affect anything other than the output of another layer.

Whatever the internal details of their architectures, these simple behaviour-based robots, consisting of several relatively independent behaviour-generating subsystems or modules, were spectacularly successful in demonstrating an ability to get around in the real world, and to carry out sequences of actions that made them appear capable of 'pursuing goals'. They turned out to be flexible, coping with unconstrained dynamic environments, and were often claimed to be robust to mechanical or electronic failures. They exploited many domains: as well as wheeled vehicles, there were legged robots, flying blimps, and submarines; behaviour-based robots prowled the corridors of MIT, and moved over the boulder-strewn replicas of the surfaces of other planets. In fact, it is possible to see behaviour-based principles at work in the robot 'Sojourner' which explored the surface of Mars in 1997. From the artificial intelligence perspective, these diverse robots showed that internal representations were not necessary for many basic abilities; Brooks himself used the proven abilities as a powerful argument in several effective attacks on the reason-based and representation-based AI programme (e.g. Brooks 1990, 1991b, 1997). The slogan 'the world is its own best model' (Flynn and Brooks 1989) reflected the outlook that sensing the real world was the appropriate strategy, at least for this level. Interestingly, the phrase appears to have been first used some two decades earlier, in exactly the same context: 'It is easier and cheaper to build a hardware robot to extract what information it needs from the real world than to organize and store a useful model. Crudely put, the...argument is that the most economic and efficient store of information about the real world is the real world itself' (Feigenbaum 1968). From the biological perspective, no formal case was ever made that subsumption or behaviour-based architectures corresponded in depth to the way in which animal behaviour was organized; the argument tended to rest on the engagingly animal-like behaviour of the robots, and to some extent on the insect-like physical appearance of some of the legged examples. The real value of these early behaviour-based robots was the convincing demonstration that coherent, flexible, and robust behaviour could be produced by nothing more than sensor-mediated switching between a set of relatively independent and simple modules, giving support to the idea that this could be the basis of behavioural control in animals with simple nervous systems, and might also form at least the substrate of behaviour in more complex animals.

The initial success of behaviour-based robotics led to a rather unstructured investigation of the potential of simple combinations of simple sensory-motor modules for producing interesting or useful behaviour. One strand used robots to re-examine the thought experiments of Braitenberg (1984), in which sensors were connected more or less directly with effectors. Another explored different methods of combining the outputs from behavioural modules; for example, instead of a single output being chosen over the others, the outputs of different modules could be summed (Steels 1993). A number of hybrid architectures were studied, with upper layers similar to conventional artificial intelligence components and lower layers inspired by behaviour-based ideas (Connell 1992; Eustace *et al.* 1994). An intriguing variant by Mataric (1990) used an upper layer inspired by certain features of the rat's brain thought to be involved in navigation. In all of these, behavioural switches were implemented in software. Snaith and Holland (1990) took a different approach, focusing on the nature and qualities of the switching mechanism itself, from both a behavioural and an engineering point of view, and examining hardware implementations; the characteristics of such switches have recently become the focus for the analysis of possible neural mechanisms for behaviour selection in vertebrates (Prescott *et al.* 1999).

Subsumption-like architectures are attractively simple and economical. This suggests that they might emerge naturally under the selective pressure of evolution. Koza (1991) performed simulations of a simple robot in which the control system was arrived at by a form of synthetic evolution (Genetic Programming); he identified the resultant control architecture as a variant of the subsumption architecture.

In Chapter 1, we noted that '...in the unconstrained animal, a number of parallel systems compete for behavioural expression. The animal switches from one type of behaviour to another, often in complex ways...' (p8). At least some of the complexity of this switching in animals is due to internal state; as we saw in Chapter 1, it is the unobservability of this state that leads to some of the most intractable problems in the modelling of animal behaviour. In the context of robotics, the key questions are different: can the use of internal state enable better control of behaviour, and how can systems using internal state be designed? Within an animal there may be several different kinds of internal state, ranging from long- and short-term learning effects, through short-term effects on internal variables (such as arousal, decay, or fatigue), to varying drive levels. As in NEWTSEX IV (see Chapter 1), analysis has often centred on factors other than learning. However, in robots, most research to date has concentrated on learning, and there have been only a small number of examinations of the potential utility of internal variables and drive levels within the behaviour-based paradigm.

Some of the most interesting ideas in this area have been examined principally in simulation. A concept common to many schemes is the idea of activation – the degree to which a module is excited. Typically, activation can be produced or modulated in three different ways: by particular sensory inputs, separately or in combination; by the action of drives analogous to thirst or hunger; and by inputs from other modules. An arbitration scheme selects the output from one or more modules, and gates it through to the system's effectors; other things being equal, the module with the highest level of activation will tend to win out over the competition.

One of the most comprehensive modular systems along these lines is Maes' behaviour network scheme (Maes 1991b), which is worth describing briefly. At any time, a module controlling a behaviour may be in either of two states: executable, when all the sensory inputs that it requires in order to be successfully executed are present; or non-executable, when one or more of the required sensory inputs is absent. A module can be directly activated by the presence of any sensor data that it requires in order to become executable, and by any 'motivation' or 'goal' with which it is associated. A module passes activation to other modules over three different kinds of directional links: predecessor links, successor links, and conflictor links. If a module is non-executable, it passes activation along a predecessor link to modules that, if executed, would cause one or more of the required sensory inputs to appear. For every predecessor link, there is a parallel successor link running in the opposite direction; an executable module passes activation along these successor links. Finally, whatever state a module is in, it passes negative activation along conflictor links to any module that, if executed, would remove any of the currently present sensory inputs required for the first module to execute. The activation that a module passes to another is always in proportion to its own activation.

When such a behaviour network is started, activation is generated within one or more modules, and spreads through the network along the links. Eventually, the activation within some executable module may rise to exceed a 'global threshold', and the module will execute, setting its own activation to zero, and inhibiting the simultaneous execution of any other module. Maes claimed that: 'When these networks are run, paths of highly activated nodes emerge, linking the sensor data to the motivations (goals). The resulting behaviour selection has been shown to be data-driven (opportunistic), goal-driven (the more a behaviour contributes to goals, the higher its activation level), sensitive to goal conflicts and it has a certain inertia or bias (towards previous behaviour sequences). I also argued...that this scheme produces resulting behaviour that is very "animal-like".' (Maes 1991c p50). Although lacking in formal proofs of convergence and stability, and in spite of a shortage of published studies of realistic simulations, and of satisfactory robotic implementations, there is little doubt that the spreading activation component within Maes' scheme may have captured some aspect of the mechanisms underlying animal behaviour in a form suitable for exploitation on robots. (See Case Study 4 for an application of Maes' ideas to frog biology.)

We can summarize the main contributions of behaviour-based robotics to the evolution of animal-like robots as follows:
1. Robots using sets of simple sensor-cued behavioural modules, with simple schemes for selecting which modules are active or controlling behaviour at any time, have been shown to produce effective behaviour sequences that strike the observer as being animal-like.
2. Robot experiments have shown that distinct observed behaviours can be produced in the absence of any module responsible for producing those behaviours, via the interaction of modules or components responsible for producing other behaviours.
3. Robot experiments have confirmed Braitenberg's thought experiments by showing that very simple connections between simple sensors and simple effectors can produce useful behaviour that looks animal-like and invites description in animal-like terms.

Although, in hindsight, all of these observations could have been made on the strength of Grey Walter's work alone, the principled framework and generality of application of behaviour-based robotics have gone a long way towards giving the biologist a flexible and accessible tool for modelling many diverse behaviours.

Robots using biologically-inspired technologies

A further influence on the development of animal-like robots has been the increasing use of technologies or components directly or indirectly inspired by biological examples. These include sensors, control systems, actuators, morphologies, and design techniques. In some cases – for example, when neural networks are used purely for their computational capabilities – the biologically-inspired element is so abstract that the robot may not seem animal-like in any respect. In others – for example, in a fast-moving robot equipped with an electronic analogue of a fly's motion-sensitive compound eye (Franceschini *et al.* 1992) – the animal connection is strong, even though the robot is far from being an artificial fly. Robots of the second type, in which some element is closely modelled on a biological system, are often called 'biomimetic robots'; several of the case studies in this volume involve such robots, and the January 2000 issue of the journal *Robotics and Autonomous Systems* was devoted to the topic. Many such robots make use of techniques derived from behaviour-based robotics.

The motivation for building a biomimetic robot may range from using it to study the biological system being modelled, to developing a robotic system with greater capability than a conventional system. It is not unknown for work undertaken for biological reasons to become diverted into the development of robot technology. An excellent example of this is the work of Lambrinos *et al.* (1997, 2000). Inspired by the navigational abilities of the desert ant *Cataglyphis*, they constructed a robot, the Sahabot, that used a polarized-light sun compass modelled on that of the ant. As well as giving insight into the structure and performance of the ant's system, the robot proved to be so accurate at navigating in the featureless Sahara that the sun compass is now under active development as a technology in its own right.

The use of biologically-inspired control systems and design techniques is a particularly popular area. The control systems take the form of analogues of nervous system components. These may consist of structured or unstructured assemblies of neurone-like elements, or larger-scale models of structural or functional modules. The design techniques may involve some form of learning, often at the level of the model neurone, or some form of artificial evolution, or some combination of the two. One reason for the interest in these approaches lies in their accessibility: because there is no intrinsic requirement for any particular sensory or motor capability, almost any simple generic robot can serve as an adequate platform, as long as it possesses adequate computational resources to implement the control system and/or the design technique. Much of this research is carried out on a single type of miniature mobile robot, the Khepera (manufactured by K-Team A.G., Lausanne, Switzerland). The Khepera (Mondada *et al.* 1993) is cylindrical, 55 mm in diameter, with two independently driven wheels at opposite ends of a diameter giving good manoeuvrability. The onboard microcontroller, a Motorola 68331, is powerful yet energy efficient. The basic sensory equipment consists of eight infra-red proximity and ambient light sensors (six at the front and two at the rear); there is provision to add further sensors and a motorized gripper. An example of the use of a Khepera to model a control system inspired by a neural model can be seen in Case Study 2 in this volume.

The use of simulated evolution to design robot control systems is almost always combined with a control system framework based on model neurones (Husbands and Harvey 1992; Husbands *et al.* 1993; Floreano and Mondada 1994; Harvey *et al.* 1994; Nolfi *et al.* 1994). Sometimes the neurones will be capable of supporting learning (Floreano and Mondada 1996). The evolutionary process is driven by the performance of the robot, but acts to alter the parameters of the neural model mediating the performance. The model neurones are much simpler than natural neurones, but retain enough of the distinctive characteristics of neurones to be of interest to biologists as well as roboticists. The tasks studied within evolutionary robotics tend to be rather simple, as do the environments, and the numbers of neurones in the control systems are typically small, of the order of a dozen or so. It has often been found possible to evolve neural controllers that achieve the desired task, but many thousands of trials may be necessary. Since robots are constrained to operate in real time, this work can consume enormous amounts of effort. One solution to this problem is to speed things up by using a robot simulation running on a powerful computer. Again, the Khepera scores: its simplicity and consistency make it possible to write simulations that correspond well with the behaviour of the real robot. The inevitable disparities introduced by noise and variability in sensors and actuators have been dealt with by the development of techniques for adding very large amounts of noise to the simulation while carrying out the evolutionary steps. Any solutions that survive these high noise levels are likely to work successfully under conditions of lower noise, such as could be expected from a real robot in the real world, and tests have found this to be the case (Jakobi *et al.* 1995).

Of course, real animals do not evolve their control systems in isolation; rather, they co-evolve them along with their sensors and effectors. This process has been studied in simulation for some time, following the fascinating work of Sims (1994). However, a recent development (Lipson and Pollack 2000) raises the prospect of being able to bring this process closer to the real world and real robots. Working in simulation, they co-evolved simple neural network controllers with simple jointed body structures and actuators, selecting for those configurations able to move the greatest distance from the start location in a given time. Using a commercial rapid prototyping technology, they then automatically constructed the bodies of the successful 'creatures', fitted the stepper-motor actuators, and connected the motors to a microprocessor running the appropriate neural network controller. The resultant constructions 'faithfully reproduced their virtual ancestors' behaviour in reality' (p976). The incorporation into this technique of some sensors would increase the computational demands of the simulation process, but would produce true robots that had co-evolved all their components, and that were therefore animal-like in respect of the processes which created them. In a parallel paper (Brooks 2000) Rodney Brooks points out another animal-like quality of this work: '...these particular robots cannot be built by conventional manufacturing techniques. The rapid-prototyping technology solidifies polymers in place, so that ball and socket joints are constructed with the ball already inside the socket. The parts are never separate, and if they were they could not be assembled without damaging them. This is not far removed from the way that biological systems grow.' (p947).

Animal–robot hybrids

A radical approach to building an animal-like robot is simply to incorporate part of an animal into a robot. At present, the motivation for doing this is the opportunity for studying the relevant part of the animal; given the difficulties of maintaining function in an isolated body part, this is unlikely to change in the near future. One of the first such systems was developed by Kuwana *et al.* (1996). Like many moths, the male silkworm moth, *Bombyx muri*, follows a trail of pheromone (volatile hormone) to locate a mate, detecting the pheromone with specialized structures on its antennae. However, rather than flying up the pheromone plume like other moths (see Figure 3.3), it follows the plume by walking. Kuwana *et al.* removed the antennae from male silkworm moths, and mounted them on a small robot. The electrical signals from the antennae were amplified and input to the robot's microcontroller, and female silkworm moth pheromone was released from a syringe in front of the robot. Two different control programs were tried on the robot: a simple Braitenberg machine, and a hand-crafted neural network. Both enabled the robot to track the pheromone successfully.

A more recent approach (Fleming *et al.* 2000) explored a different type of hybrid. The nervous system of the anaesthetized larva of a sea lamprey (*Petromyzon marinus*) was dissected out, and connected electronically to a Khepera so that the input from the Khepera's light sensors stimulated the vestibular centres of the preparation, and the output from the preparation's motor centres drove the Khepera's motors. Under the control of the lamprey's brain, the Khepera produced a variety of behaviours in relation to lights in the environment, showing patterns of either attraction or repulsion. The experimenters went on to study changes in these patterns caused by neural changes in response to sustained exposure to light coming from only one side; they were able to explain the subsequent changes in behaviour as neural adaptation to the input conditions.

Artificial ethology

The case studies in this volume draw on many of the ideas discussed in this chapter, and have been selected to show how real and simulated robots have a distinctive contribution to make in understanding the nature and causes of behaviour in natural and artificial systems. Each set of case studies is set in context by supporting material from ethology and psychology. We believe that robotics can be relevant to ethology, and also that ethology can be relevant to robotics, and we hope that this collection may serve as an effective introduction to artificial ethology, the extension and integration of the two disciplines.

Chapter 3

Sensory processes and orientation

Before Muller's (1827) doctrine of specific nerve energies it was thought that each specific environmental event, or stimulus, affected sensory nerves in a manner specific to the type of stimulus. Muller discovered that a given nerve pathway always produces the same type of sensation regardless of its mode of stimulation. Thus light falling upon the eye produces a visual sensation, but so does mechanical stimulation, such as a blow to the eye. If our ear could be attached to the optic nerve, auditory stimulation would then give rise to visual sensation. The sense organ acts as an energy transducer providing quantitative information. The destination of the nerves that relay the sensory messages determines the qualitative information. The doctrine of specific nerve energies provides a key organizing principle for sensory physiology. It banishes all previous speculations as to the role of sense organs in perception, making it clear that these are primarily transducers of one form of energy to another.

Sensory receptors are specialized nerve cells responsible for the transduction and transmission of information. Like ordinary nerve cells, they have a cell body, dendrites, and one or more axons. Receptors are specialized according to the type of environmental energy to which they react. For example, photoreceptors contain pigments that are chemically modified by light and that give rise to an electrical potential when this occurs. Mechanoreceptors undergo electrochemical changes as a consequence of deformation of the cell membrane. This transduction of energy usually takes place in the cell body, and a characteristic of all receptor cells is that the environmental energy is converted into a graded electrical potential, called the generator potential. This potential is usually proportional to the intensity of stimulation of the receptor. When the generator potential reaches a certain threshold level, it triggers an action potential that travels along the axon of the receptor cell. This is the transmission part of the sensory process, and the information is usually coded such that the more intense the stimulus, the higher the frequency of action potentials.

The action potentials conveying sensory information are no different to any other nerve impulses. Their magnitude is determined by the size of the neuronal axon and their frequency by the intensity of stimulation. Each type of receptor sends impulses, either directly or indirectly, to a particular part of the brain. The sensations experienced depend not upon the type of receptor or the messages they send but upon the part of the brain that receives the message. The brain is also responsible for the localization of the sensation, either within, or outside, the body. In the case of pain, for example, the nerve fibres from the hand go to one part of the brain, those from the arm go to another and so on. The pain experienced by the brain is referred to that part of the body from which the message came. In the case of external stimuli, localization of the stimulus depends upon the anatomical arrangement of the sensory receptors.

The orientation of the whole animal in space depends upon the localization of external stimuli. It may be based on very simple principles but may also involve complex mechanisms. The simple principles can been seen most easily in certain invertebrate species. Gottfried Fraenkel and Donald Gunn (1940) proposed a system of classification based on the work of earlier writers. This system has provided the basis of more recent discussions and reviews (for example Kennedy 1945; Adler 1971; Holland and Melhuish 1996).

The simplest form of spatial orientation is kinesis, in which the animal's response is proportional to the intensity of stimulation but is independent of the spatial properties of the stimulus. For example, common woodlice (*Porcellio scaber*) tend to aggregate in damp places beneath rocks and fallen logs. They move about actively at low humidity levels but are less active at high humidity levels. They consequently spend more time in damp conditions, and their high activity in dry conditions increases the chances that they will discover a damp place. Similar behaviour is shown by the ammocoete larva of the lamprey, which varies its swimming activity in accordance with the light intensity (Jones 1955).

The type of kinesis in which a relationship exists between the speed of locomotion and the intensity of stimulation is called orthokinesis. Another type of kinesis, shown by the flatworm *Dendrocoelum lacteum*, is klinokinesis. Here the rate of change of direction increases as the light intensity increases (Ullyott 1936; Hinde 1970).

In a number of types of orientation, usually grouped together as taxes, the animal heads directly towards or away from the source of stimulation. For example, when the larva of the housefly (*Musca domestica*) has finished feeding, it seeks out a dark place where it pupates. At this stage it will crawl directly away from a light source, and it is said to show negative phototaxis. The maggot has primitive eyes on its head that are capable of registering changes in light intensity but that are not able to provide information about the direction of the light source. As the maggot crawls it moves its head from side to side. When the light on the left is brighter than that on the right, the maggot is less likely to turn its head towards the left. It thus tends to crawl more towards the right, away from the light source. In response to an increase in the level of illumination, the maggot increases its rate of head turning. If an overhead light is turned off every

44 *Artificial ethology*

time the maggot turns it head to the right and is turned on every time it turns to the left, then the maggot turns away from the illuminated side, describing a circle towards the right. Thus, although the animal has no directional receptors, it can perform a directional response. Similar behaviour is shown by the single-eyed organism *Euglena* (Fraenkel and Gunn 1940, 1961). What can be achieved by an agent with a single sensor has been explored by Holland and Melhuish (1996).

Orientation by successive comparison of stimulus intensity requires turning movements. Usually this is called klinotaxis. Many animals show klinotaxis in response to gradients of chemical stimulation. Simultaneous comparison of the intensity of stimulation received at two or more receptors enables the animal to strike a balance between them. It can then achieve tropotaxis that enables it to steer a course directly towards or away from the source of stimulation. For example, the pill woodlouse (*Armadillidium vulgare*), which lives under stones or fallen trees, shows a positive phototaxis after periods of desiccation or starvation. With its two compound eyes on the head, the animal is able to move directly towards a light source. When one eye is blacked out, however, it moves in a circle. This shows that the two eyes normally provide a balance of stimulation. When presented with two light sources, the woodlouse often starts off by taking a median course but then heads towards one light. This happens because equal stimulation of the two eyes can be achieved either by steering between two sources or by moving directly towards one. Deviations from one source tend to be self-correcting, and lateral sources of light tend to be ignored because the eyes are shielded at the back and sides. Deviations from two sources set wide apart, therefore, may result in contact with one being lost.

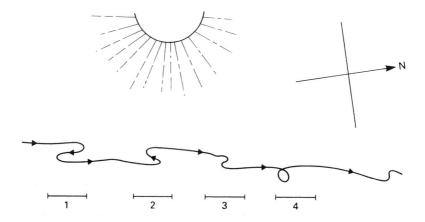

Figure 3.1. *Sun compass orientation in ants.* In Santschi's (1911) classic experiment the ant was shaded from the sun and sunlight was reflected onto the ant from the opposite direction. When this was done at positions 1 – 4 the ant changed direction, as illustrated. (*A*fter Schöne 1984.)

Eyes that are capable of providing information about the direction of light by virtue of their structure are capable of telotaxis. This is a form of directional orientation that does not depend upon simultaneous comparison of the stimulation from two receptors. When there are two sources of stimulation, the animal moves towards one and never in a median direction. This shows that the influence of one of the stimuli is inhibited.

Menotaxis is a form of telotaxis (Hinde 1970) that involves orientation at an angle to the direction of stimulation. An example is the light compass response shown by homing ants. These animals are guided, in part, by the direction of the sun. Santschi (1911) shielded an ant trail from direct sunlight and used mirrors to reflect sunlight onto the trail from the opposite direction. Ants entering the region illuminated by reflected light turned around and walked in the opposite direction, as shown in Figure 3.1.

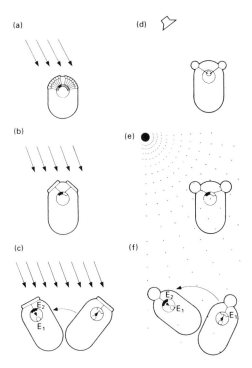

Figure 3.2. *Schematic representation of some of the basic principles of sensory orientation.*

(a) The direction of stimulation (e.g. light) is registered by a raster of sensory receptors. (b) Direction registered by simultaneous comparison between two receptors. (c) Only one receptor is available and the animal makes successive comparisons by moving its body. (d) Time of arrival of stimulation (e.g. sound waves) is compared by two receptors. (e) A gradient of stimulation (e.g. chemical) is registered by comparison between two receptors. (f) A gradient is registered by a single receptor as the animal moves to sample different localities. (From *The Oxford Companion to Animal Behaviour* 1981.)

The type of orientation that can be achieved in a given situation depends jointly upon the nature of the external cues and the sensory equipment of the animal. An animal with only one sensor that is sensitive to stimulus strength only is limited to successive measurements of stimulus strength in different localities. If the external cues are inherently directional, then a single receptor that is shielded on one side can provide directional information. Thus, a shielded photoreceptor is useful in this respect, but a shielded chemoreceptor is of no advantage because chemical stimuli are not inherently directional. With two receptors, simultaneous comparison can be used to detect gradients (see Figure 3.2). With many receptors arranged in the form of a raster (a row or mosaic arrangement), more sophisticated types of orientations can be achieved. Examples of rasters are the lens eyes of vertebrates and the compound eyes of arthropods.

Spatial orientation is often achieved by a combination of methods. For example, some moths are attracted to females as a result of the airborne pheromone released by the female. The scent is windborne, and the flying moth must orient with respect to the wind. Flying animals normally use visual cues to monitor their progress with respect to the ground. The flight path of the animal is affected by the wind direction, giving a resultant track (Figure 3.3). Experiments with moths show that the track angle changes with the scent concentration. When the scent is absent, the animal flies backward and forward without progressing upwind (that is, with a track angle of 90°). When scent is

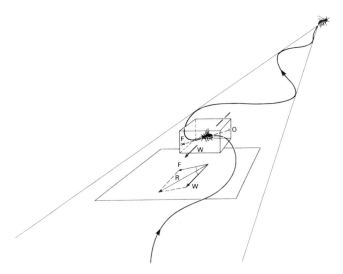

Figure 3.3. *Male moth flying upwind in response to pheromone released by a female.* The flight path, F, is affected by the direction of the wind, W, giving a resultant track, R. The track angle is the angle between the wind direction, W, and the flight direction, R, with respect to the ground.

detected in the wind, the track angle increases and the animal zigzags upwind. The changes of direction are related to the borders of the scent trail, as illustrated in Figure 3.3. When the scent concentration drops below a certain level, as at the edge of the scent plume, the animal turns in a direction opposite to that of its previous turn. This turning behaviour is not related to wind direction, but it depends upon internal reference, or idiothetic information. Thus, the flying moth uses a combination of visual, amenotaxic (wind), and idiothetic orientation mechanisms in searching for a mate.

One might think that chemical orientation in relation to water currents is similar to that in air. Recent studies show that there are important differences, and that the Fraenkel and Gunn classification of animal orientation is not as all embracing as it is usually thought to be.

Case study 1: How robotic lobsters locate odour sources in turbulent water *by Frank Grasso*

Lobster chemotaxis

Robolobster was constructed as an aid to understanding how American lobsters (*Homarus americanus*) locate odour sources in a turbulent marine environment. The traditional framework for understanding animal orientation problems, as exemplified by Fraenkel and Gunn (1940) invokes kineses, klinotaxes, and tropotaxes as generic mechanisms independent of the individual species context. The major limitation of these frameworks that concerns us here is that, when they are applied to understanding a specific problem of chemo-orientation, they assume a uniform gradient of the chemical that may be tracked through space. This is not a problem for creatures like nematodes which appear able to use simple algorithms that assume a smooth chemical gradient produced by diffusion as the dominant dispersal mechanism (Ferrée *et al.* 1997; Morse *et al.* 1998). For creatures like lobsters, which live in environments where inertial rather than viscous forces dominate chemical distribution patterns, smooth gradients of chemical signals between the odour source and the animal's sense organs are the exception rather than the norm (Moore and Atema 1991; Dittmer *et al.* 1995, 1996; Grasso *et al.* 1998).

When, in the laboratory, we examined the physics of the odour distribution under conditions that reproduced the pattern of chemical dispersal downstream from a clam's feeding current (Monismith *et al.* 1990) we found that the 'odour' would make its way to a hungry lobster's chemosensors as a series of odour patches in space rather than a smooth concentration gradient (Moore and Atema 1991; Dittmer *et al.* 1995; Basil *et al.* 1998). Yet, American lobsters orient successfully to these laboratory sources (Moore *et al.* 1991; Basil and Atema 1994) as they (presumably) do to clams in the field. Somehow they are able to

48 Artificial ethology

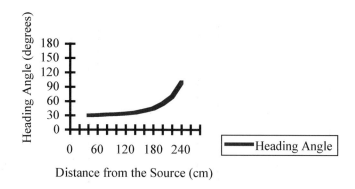

Figure CS 1.1. *Lobster heading angle during chemo-orientation.* Lobsters systematically improve their heading angle to the odour source during chemotaxis. Heading angle is defined as the angle (or error) between the animal's direction of motion and a direct path from the animal's location to the source.

bridge the gaps between odour pulses and track to the odour source in the absence of a concentration gradient, and simple time averaging is inadequate (Grasso *et al.* 1998). Indeed, Moore *et al.* (1991) report that the error in the animal's direction of travel compared to the direct path to the source decreases as the lobster has greater experience of the plume (Figure CS 1.1). This suggested that the animal's successful odour-tracking episodes were not a matter of chance but the result of a sophisticated set of mechanisms that could interpret the patterns of odour produced by turbulence. There are various ways in which this could be achieved (Figure CS 1.2).

Robolobster

At the Marine Biological Laboratory, Woods Hole, we set about, and are in the midst of, a three-part programme of research to try to understand how the lobster accomplishes this chemical source localization. Until recently we have used a single, reproducible, fluid-dynamic regime we call the standard plume (Dittmer *et al.* 1995), in which we studied the physics of odour dispersal, the chemo-orientation behaviour of lobsters, and the chemo-orientation behaviour of our lobster biomimic, Robolobster. I believe that the interplay between these three approaches has led us to design experiments that are more directed towards the overall goal of understanding lobster chemo-orientation than any of these approaches alone would have permitted. The certainty that comes from being able to say 'this is what the lobster does' or 'the plume generates signals of this kind' or that 'the robot tells us that that just won't work' has allowed us to select experiments that will provide us the greatest amount of new insight.

Robolobster (Figure CS 1.3) was built to approximate the scale of the American lobster in size, speed, manoeuvrability, and olfactory sampling. (The

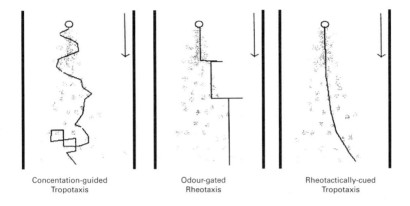

Figure CS 1.2. *Idealized paths from three plume-tracking algorithms.* The source is marked with a circle; the mean flow is in the direction of the arrow.

details of Robolobster's construction can be found in Consi *et al.* 1994, 1995.) We did not aim to mimic lobster biomechanics (but see Ayers *et al.* [1992], Crisman and Ayers [1992], and Ayers in this volume for an example of a biomechanical lobster biomimic). Rather, we created a framework that allowed us to study the functional and dynamic relationships of plume orientation on the scale of an orienting lobster. Robolobster is equipped with a pair of conductivity sensors, which possess a temporal integration time that is averaged to match lobster chemo-receptor dynamics. Robolobster is propelled by two wheels, which receive power under independent control. This allows us to match the speed and turning of the robot to that of the orienting lobster. Robolobster is completely autonomous in underwater operation, transforming sensor inputs into motor commands under the control of programs loaded into the memory of a TattleTale Model 7 computer. For analysis of the behaviour of the robot, we log the sensor inputs, monitor the path of the robot with video, and calculate the turning behaviour of the operating robot with the integrated signal of an onboard gyroscope.

Our strategy with the robot has been to work from simple, to more sophisticated, algorithms, modifying the control algorithms as necessary to improve performance and to test hypotheses of lobster chemo-orientation. Our main focus and interest has been the sensory processes involved in interpreting chemical signals. Because, as mentioned above, the American lobster appeared to be tracking odour plumes using a method other than pure chemotaxis (see Moore *et al.* 1991; Basil and Atema 1994), our first goal was to determine how effective pure chemotaxis (i.e. guidance from chemical cues alone) can be in tracking turbulent odour plumes to their source. In these studies, we programmed Robolobster to respond to neutrally buoyant salt plumes (see Dittmer *et al.* 1995) in a laminar freshwater background flow.

Figure CS 1.3. *Photograph of Robolobster in a flume with an example of its namesake in the foreground.*

Robolobster senses salt concentration with a pair of conductivity sensors (measuring in microSiemens — µS) mounted on the front of the robot. We treated the robot as if it were an animal, scoring its performance in repeated trials under the same conditions and with the same measures as applied to the analysis of lobster chemotaxis. To date we have examined only two simple algorithms. In these trials the robot starts in the plume (it is not required to locate the plume) and moves continuously forward at 9 cm/sec (lobster tracking speed) unless conditions trigger the invocation of one of the following two rules:

1. Robolobster turns towards the side of the higher salt concentration signal or moves forward if the intersensor difference drops below 9 µS.

2. Robolobster moves backwards if both sensors detect background concentrations (freshwater) (both sensors register < 7 µS).

Algorithm 1 employs Rule 1. Algorithm 2 implements both rules.

In the following pages I will not present our experiments in any great detail. These may be found in papers that have already appeared in print. Instead I will illustrate, with examples drawn from our work, the potential of biomimic research for shedding light on areas of interest to the ethologist.

Direct tests of explicit hypotheses

The greatest advantage that an ethologist can derive from the application of biomimetic robots is to have the opportunity to test explicit hypotheses of mechanisms of animal behaviour under the conditions that the animal encounters. Naturally, if the hypothesis, implemented in a robot, reproduces the animal behaviour it does not prove that the hypothetical mechanism is the one that the animal employs. Robots can provide a proof that the mechanism is competent under the same conditions that the animal encounters. Of perhaps greater value is that a failure of such a mechanism when tested in a biomimic indicates that that concept of the behaviour (perhaps long cherished in the literature) simply cannot work under the conditions the animal confronts. The

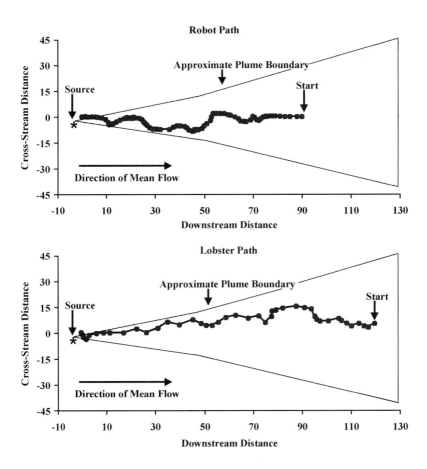

Figure CS 1.4. *Typical Robolobster and lobster paths.* The robot's path (upper panel) contrasts with that of the lobster (lower panel). Robolobster, running Algorithm 1, makes a series of sweeping turns as it approaches the source (marked *).

ability to exclude such hypotheses can delineate unproductive avenues of research for the ethologist interested in understanding a particular animal behaviour.

With Robolobster, we examined pure chemical tropotaxis (the instantaneous estimate of a concentration gradient with paired sensors) and determined that such mechanisms are effective at only short distances from the source of our standard plume. Furthermore, the paths generated by the robot, when controlled by these algorithms, did not resemble quantitatively or qualitatively those of actual lobsters under the same conditions. The results make the hypothesis of a concentration-based tropotaxis exceedingly unlikely as an explanation of lobster chemotaxis (Figures CS 1.4 and 1.5).

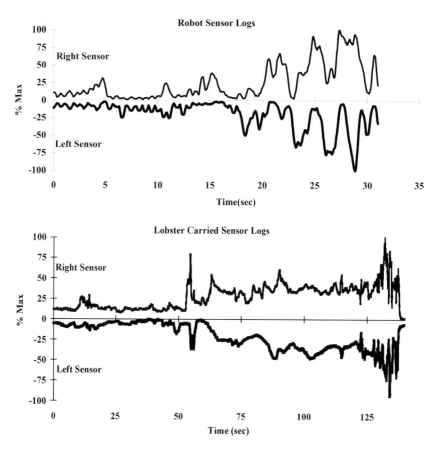

Figure CS 1.5. *Typical Robolobster and lobster concentration profiles.* The upper panel shows salt concentration signals taken from the robot's sensory logs. The lower panel shows dopamine concentration profiles recorded with a pair of electrochemical electrodes mounted on the back of the lobster and positioned in the olfactory sampling area of the lobster. Dopamine was used as a tracer in the food odour plume the lobster tracked. Both sets of curves are expressed as a percentage of the maximum signal. Robolobster's counterturning near the source is evident in the sensory log. It is clear that the lobster is not using this algorithm.

The underlying assumption of these studies is that when the lobster tracks an odour plume it is simply invoking a rule or (more likely) a set of rules and that therefore its behaviour is controlled by the stimulus patterns produced by the plume. Thus we learn that the better of the two algorithms is successful on about 30% of the trials over a distance of 100 cm. Yet, direct comparison of the success rates of lobster and Robolobster is more complicated than comparison of path characteristics. We imagine that the lobster's performance is 'near perfect' but of course the animal's behaviour is not entirely governed by the patterns of stimulus it receives through antennular chemoreceptors: it has other sources of information and may not always be motivated to track the odour plume. These considerations lead to a number of alternative questions which begin to take the shape of more sophisticated hypotheses.

We cannot determine whether the lobster's non-successful (by our definition) trials are due to an absence of motivation, or a distraction of the animal, or whether the lobster is waiting for conditions under which a tropotactic algorithm would be successful. These alternatives (and they are not the only ones) raise issues of the complexity of the model. Regardless of which approach to enhancing the robot model is taken, the added complexity of the next generation of hypotheses is justified by the failure of the simple hypothesis in this critical test.

Contribution of specific modalities can be isolated for study

Animals rarely rely on just one sensory modality at a time in the execution of a given behaviour. This provides the animal with a great deal of behavioural flexibility and the ethologist with the thorny problem of deciding which sense is being used at which time in the execution of a given behaviour. The problems of interpreting the results of lesion (tissue removal) experiments even where complete removals are possible are well known. The ability of animals to compensate for such losses makes the validity of observed behaviour for the intact animal a difficult matter to establish. Robolobster has allowed us the ability to study the effectiveness of one modality in perfect isolation to learn what is possible with that sense in the lobster's context: the robot cannot compensate unless we specifically program it to do so.

This fact gives us a reasonable degree of certainty that the lobster is not able to use simple concentration signals alone for guidance to the odour source. The lobster may be doing some more sophisticated analyses of the chemical concentration profile and our studies of the physics of odour dispersal suggest workable alternative chemical signal processing strategies. When we have tested these we will be able to ascertain, as we did with Algorithms 1 and 2, the degree to which these new algorithms produce paths that are similar to those of the lobster.

Of still greater importance is the fact that we are in a sense mapping the limitations of a particular modality, olfaction, from the lobster's point of view. Each tested algorithm tells us what can and cannot be a role fulfilled by the chemical sense of the animal based on an understanding of the stimulus. This practical knowledge base will allow us (hopefully) to be better able to integrate multiple sensor modalities (flow for example) into the robot as they become necessary. Each modality studied in context will allow us to know what we can expect from each individually before we attempt their integration.

Identify the limitations of specific modalities

The complement of asking what it is that a particular modality can contribute to a particular behaviour is asking what it is that that modality cannot provide to the animal. The identification of the limitations of a modality is crucial for hypotheses dealing with the contribution of a number of sense modalities to informing the behaviour. In this context the questions the robotocist must ask are different, in a sense more practical, than those of the experimenter: not 'what modality is the animal using to get this information ?' but rather 'what information is available from this modality for this task ?'.

Our experience with Algorithm 2 made part of this clear to us the instant we began running the program in an actual plume. We knew from our previous studies of the physics of odour distribution that the odour distribution was discontinuous in space in the standard plume. We programmed the robot to average these signals on a scale that was consistent with what we know of the dynamics of lobster chemoreceptor response and we thought that this, combined with the motion of the robot through the plume (at lobster walking speed) might be enough to smooth these discontinuities and allow the gradient to be sensed. The frequent direction reversals by the robot running Algorithm 2 provided us with visible evidence every time a discontinuity was reached, and showed us that this was not the case.

Without the use of Robolobster this kind of hypothesis becomes virtually impossible to test. This is largely because existing theoretical models of fluid flow are not able to capture the action of turbulence on our odour signal at two or more points at the 250 ms time scale. But even if these calculations could be converted into a simulation we would still require a physical plant model of the robot and its effect on the plume as it moved through it to answer this question. At that stage it becomes far simpler and direct simply to run the robot.

Advantages over pure simulation studies

Megan Mahaffy, Christian Reilly, and myself have studied a number of alternative chemo-orientation strategies in simulation. The advantage of these studies is that one can compare a large number of alternative strategies in a very short period of time (Holland and Melhuish 1996). The extent to which these results can be applied rests with the accuracy of the simulated environment in which the animat behaves.

In our work we constructed an interpolated two-dimensional spatial map of the statistics of odour distribution based on actual measurements taken from an instance of the standard plume. We then tested a variety of single sensor algorithms (two sensor algorithms were not possible) and learned something about the rate at which useful guidance information could be obtained by each. The algorithms we examined were klinotactic, steering into the gradient detected in a series of sequential samples. We varied the form of the information (the cue from which the animat derived its guidance) from pure concentration to a variety of extracted temporal parameters. We found that a number of parameters, particularly the onset slopes of the peaks of the concentration signal, led the animat to the simulated source in an order of magnitude less time than pure concentration (Figure CS 1.6).

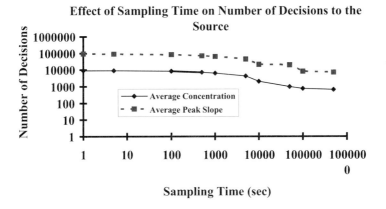

Figure CS 1.6. *Relative success of klinotactic mechanisms in simulation.* This figure plots the average number of decisions an animat must make ($n=100$) to reach the source as a function of the time the animat spends at each sampling location making estimates of the local concentration or average peak slope. The lobster makes decisions on the 2-4 sec time scale so it is clear that these mechanisms are not those that the lobster uses. This figure points out the fact that other parameters of the odour signal besides concentration can make information available to the animal at a much greater rate, in this case, and order of magnitude.

These results provide us with a starting point, and an easy early screen for evaluating multiple hypotheses. On the other hand this is clearly not the same thing as running the actual animal or the robot. The robot running an algorithm cued by one of these parameters would affect the plume downstream from its path simply by moving through it suggesting that the actual efficiency would be unpredictably lower. Also, for turbulent odour plumes the ability to produce single point statistics, as in our interpolated playback model, is a straightforward matter. The models for producing joint simultaneous statistical models for pairs of points at the subsecond time scale are not available. This difficulty is multiplied when one considers developing simultaneous environmental models to inform multiple sensory modalities in an animat. The problems of synchronization alone, when considering local flow cues combined with chemical cues for modelling odour plumes, raise serious issues of the model's validity. With biomimic studies in context we find two advantages compared to simulation studies. Firstly, the ethologist in search of answers halves the conceptual problems by concerning himself or herself with only the validity of the biomimic, rather than the biomimic and its context. And secondly, by running the biomimic in the animal's context he or she has the opportunity to study the animal's sensory world directly.

Exploring the stimulus world of a given animal from the animal's viewpoint

When we used the chemo-tropotactic algorithms in tracking the standard plume we recognized two distinct patterns of the Robolobster paths under the second algorithm. At distances far from the odour source the robot's path was marked by many turns and direction reversals; closer to the source the paths showed smooth high radius turns and no direction reversals. This was directly traceable to the patchiness of the plume at greater distances. Nearer the source the stimulus jet's momentum kept the plume tightly together providing a sharp plume edge for the robot to follow.

In this case the performance of the robot outlined to us the physical location of a transition in plume quality. Success of either algorithm was quite high when the robot started in the 'proximal jet' but Algorithm 1 was hopeless at distances greater than 50 cm. When the robot running Algorithm 1 exits the plume it can receive no signal that would allow it to return to the plume and it continues to move away. Algorithm 2 achieved a 30% success rate when it started in the 'distal patch field' primarily because the direction reversal instruction kept it in the plume long enough to wander to the proximal jet where edge tracking of the plume is possible.

It is interesting to note that the lobster too undergoes transitions in behaviour at different distances from the source (Moore *et al.* 1991). At distances below 30 cm from the source the lobster's path is more irregular. It is thought that the behaviour at this stage involves the lobster raking its legs across the flow to make physical contact with the clam, should it be lying on the bottom. Thus where the lobster is irregular in its path the robot is most regular and vice versa. This provides further evidence that neither of our two algorithms is a good model for the lobster's behaviour but at the same time it draws our attention to the full significance (in terms of chemical signals) of the difference in the environment encountered by the lobster at these two phases of the orientation task.

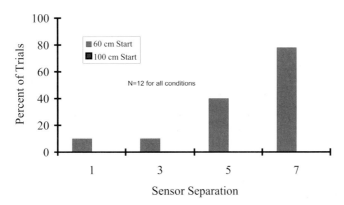

Figure CS 1.7. *Effect of sensor separation on number of hits.* Number of hits scored by Robolobster running Algorithm 2 with sensors at various separations. No hits were scored in this experiment when Robolobster started the trial 100 cm downstream from the source.

Tool for examining evolutionary alternatives

The use of robots in the animal's context give the student of animal behaviour a greater degree of flexibility in asking questions of why an animal is designed a certain way. He or she can vary some design feature of the animal and see the effect that variation has on the overall performance. By pushing a single feature to the extreme the ethologist can detect when a design is optimized for a particular strategy, when some other strategy is employed or when a trade-off (phylogenetic inertia or multiple functions for the same organ) has shaped the design. In short it provides an indirect window into evolutionary questions.

In one study with Robolobster (Figure CS 1.7) we examined the effect of sensor separation on the robot's ability to locate the source in the standard plume. The straightforward prediction for concentration-based algorithms is that if a concentration gradient exists in the plume it should be more easily detectable with greater inter-sensor distance (this is essentially the definition of gradient). We found that this condition was borne out in the region of the plume that was nearest the source (within 50 cm) but when the robot started at greater distances we did not see an improvement in the number of times Robolobster 'hit' the source.

Our odour-plume-tracking American lobsters typically hold their antennules upright, separated from one another by a distance of about 3 cm. At a 7 cm sensor separation and running Algorithm 2, Robolobster's hit rate was about 80% within a trial time of 30 seconds when the robot started 60 cm from the source and 0% when it started 100 cm from the source. This suggested to us that further downstream from the source there was no gradient to track, while nearer the source the gradient was actually the plume boundary. Robolobster paths from these two conditions confirm this: there was no statistically significant progress towards the source evident at the 100 cm starting distance with the tripled sensor separation. The frequent alternations of direction reversal, turns and forward progress by the robot provide ample evidence that the signals were well above the robot's detection threshold. Thus it would appear that edge tracking is not a useful strategy when the plume has spread in space to form many disparate patches.

This observation leads back to the lobster. If it could improve its concentration-gradient tracking performance with increased sensor separation, why should it have such a small one? Spiny lobsters, animals of comparable size and habitat, have much longer antennules (12–13 cm) that they hold at much greater separations when tracking. The American lobster, it seems, has selected another orientation strategy than tropotactic concentration gradient tracking.

Generalization

The ethologist is forced to solve the same problems evolution has posed for the animal's ancestors to produce in a robot a degree of performance comparable to that of the real animal. The objection is often raised that there is no guarantee that the solution obtained with a robot is the same as that which evolution has provided the animal. This is true: robots can only provide proofs that a given solution is workable. The question of the 'reality' of the solution must be addressed by careful comparison of animal and robot performance under identical conditions and similar manipulations. Thus, the robot adds to, but can never replace, the tool chest of an ethologist who wants to explain an animal's behaviour. Along the way, of course, alternative workable solutions to the orientation problem will be of great interest for practical robotics and for the study of evolution as viable examples of 'the path not taken'.

Acknowledgements
The majority of the results reported here resulted from collaboration on the Robolobster project. The core of the Robolobster 'team' who participated in this work included in addition to myself: Jelle Atema, Thomas Consi (the originator of the idea to construct Robolobster), and David Mountain. Dr Jennifer Basil, Jonathan Dale, Paul DiNunno, Kevin Dittmer, Diana Ma, Megan Mahaffy, Christina Manbeck, and Christian Reilly all made significant contributions to this effort during various phases of the work. Much of the work reported here was supported by NSF grant BES-9315791 and NSF grant IBN-9631665 to Jelle Atema.

End of Case Study 1

Auditory orientation

The use of robots has also been instructive in the study of auditory orientation. Male crickets attract females over long distances by a calling song. The sound is produced by rhythmic opening and closing of specialized forewings that carry friction mechanisms. The songs are remarkably stereotyped among members of a local population, but they differ considerably from one species to another. The differences occur primarily in the temporal patterning of the pulses. The female of the species has the problem of identifying the song as one of her own species, and of orienting towards, and finding, the source of the song. She may also have to choose among the songs of competing males. How this is done has been the subject of a considerable amount of research, in which robots have recently played an important part, as we see in the next case study.

Case study 2: Robotic experiments on cricket phonotaxis *by Barbara Webb*

Real crickets

Male crickets attract females by producing a calling song, which the females approach. The specific form of the song is species dependent and identifying characteristics are the frequency range and temporal patterning of the song. Many crickets produce songs with a relatively pure carrier frequency and a pattern of repeated short 'syllables' of sound. The forewings have a rasp and file such that drawing one across the other causes the wing to resonate at a certain pitch, and the pattern is set by the repeated wing movements (Figure CS 2.1). Female crickets can locate males purely by moving towards this calling song. That is, in experiments with no chemical or visual cues they can directly approach a speaker producing the sound, although in real mating behaviour they may well use other cues or search to discover males.

Females appear to identify reliably the features of conspecific calling song and approach only 'correct' sounds. More precisely it has been shown that the probability and directness with which a cricket approaches an example song changes when parameters of the song are varied. In particular, songs of higher or lower frequencies than the normal carrier frequency are not as attractive; and songs with repetition patterns faster or slower than the normal pattern are also not as attractive. The extent to which other features of the song affect taxis appears to vary between species and experimental paradigms.

Females robustly and reactively approach the sound by turning to the side where the sound is stronger. By 'robust' is meant that this strategy is very effective in approaching a sensory source even when there is substantial noise and error. By 'reactive' is meant that the cricket appears to follow this strategy with no memory of how far it has turned or travelled, continuing to walk in

60 Artificial ethology

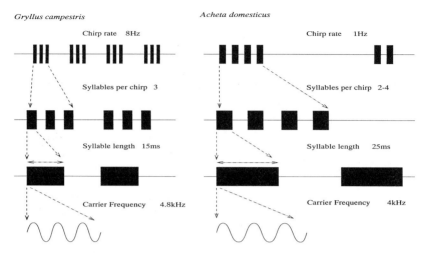

Figure CS 2.1. *Different temporal features of cricket song.* The critical parameters for species recognition appear to be the carrier frequency and the intersyllable interval (the time between syllable onsets within a chirp).

circles on a treadmill for minutes to hours. Apparently it has no representation of relative spatial position of its target but simply uses this basic strategy to get closer to it.

Females choose between simultaneous calling songs. If more than one male is singing within earshot (which is frequently the case), females do not appear to get lost or confused between them but approach one male. This may be at random if the songs are identical, or be preferential for songs with certain characteristics e.g. louder, faster or closer similarity to the correct calling song.

How does the female perform this task? Despite many years' investigation, it is still not known exactly what combination of sensory mechanisms and neural processing controls the behaviour (Figure CS 2.2). The main unresolved issues to be explored here are:

1. Does the cricket actively filter sounds for the right characteristics or are its preferences better explained as side effects of a limited ability to localize?

2. Does timing rather than frequency of spikes play a role in the neural control of this behaviour?

Robot crickets

One approach to the questions posed above is to implement a robotic mechanism based on data and hypotheses about the cricket system and to test the robot's ability to perform the cricket behaviour.

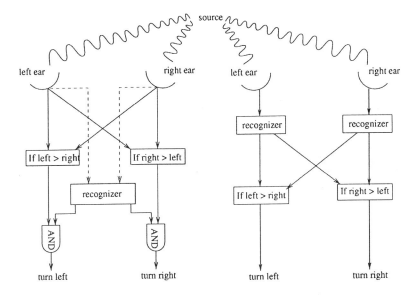

Figure CS 2.2. *Two possible mechanisms to explain cricket phonotaxis.* The system on the left uses a single recognizer and compares left-right signal strength; the system on the right compares the outputs of bilateral recognizers.

Two rather different robots have been used so far in these studies. The original experiments were carried out on a Lego-based robot, controlled by a programmable microprocessor. This system had the advantages of simplicity, but the main drawbacks were the relative size of the robot compared to the distance travelled to the sound source, the poor motor control, and the fact that the auditory circuit processed sound too slowly to use real cricket song as a stimulus. With this prototype robot, it was successfully shown that, using slowed-down song, the selectivity of cricket females for certain repetition rates could be reproduced with a relatively simple algorithm that meant the robot could not locate a sound without the right temporal parameters (see below).

The current robotic cricket is based on a Khepera miniature mobile robot. This measures roughly 6 cm in diameter and 10 cm high. There are two drive wheels and two castors. A customized auditory circuit (see below) was designed to interface with the processor, and could produce sound readings at a rate of more than 1 kHz enabling us to use real cricket song as input (Figure CS 2.3).

The cricket faces a difficult problem in determining the direction of the sound source it wishes to approach. The distance between its ears, which are on its forelegs, is small relative to the distance of the sound, and the wavelength of that sound is large relative to the body size of the animal. Consequently there is little direct amplitude difference in sound at the two ears either from distance attenuation or sound shadowing. The phase difference is also small

62 Artificial ethology

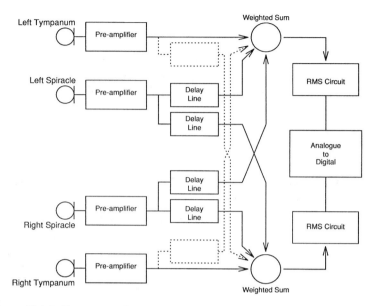

Figure CS 2.3. *Electronic implementation of auditory system.* This block diagram shows the elements and connections of the auditory circuit implemented on the robot to mimic the auditory system of the cricket.

(microseconds) and the cricket lacks the specialized neural processes found in owls and mammals for detecting this difference directly. Consequently it has evolved a unique solution where the phase difference is physically converted into an easily detected amplitude difference. The eardrums on the legs are connected by an air-filled tracheal tube to each other and to two additional openings on the cricket's body (see Figure CS 2.4). Sound thus reaches the eardrum (tympanum) both directly and indirectly, via the tube, so that its vibration reflects the summation of these different waves. The phase relationships of these waves depend on the distances travelled (and internal delays in the tube), which depend on the direction of the sound source. Thus a sound source on the left side can produce a larger vibration in the left tympanum than the right tympanum.

This mechanism was modelled electronically, using two microphones with a small separation — equal to approximately a quarter of the wavelength of the calling song. The output of each microphone was combined with the inverted, delayed output of the other microphone. The delay represents the time for sound to travel through the tracheal tube and the inversion the fact that direct and indirect sound operate on opposite sides of the eardrum. The result is that sound amplitude varies with sound direction in an approximately cardioid (heart-shaped) function. One immediately obvious feature of this mechanism for directionality is that it is frequency dependent: the phase relationship with fixed separation of the ears and fixed delays between them will vary with the

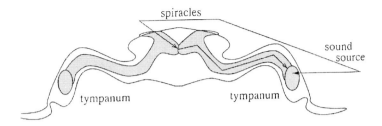

Figure CS 2.4. *Sound paths in the cricket auditory system.* The tracheal tube connecting the cricket's ears makes them into pressure difference receivers.

wavelength of the sound. A frequency of half the 'tuned' frequency will produce half as much amplitude difference from right to left, and a doubled frequency will produce no directional difference at all. Consequently, the cricket may fail to approach sound sources of the wrong frequency because it cannot localize them as accurately, rather than because they are 'recognized' as incorrect for the species. Although the cricket does have some tuning specificity in its auditory receptors, this may not be necessary for the selectivity of its behaviour for carrier frequency

The implemented circuit, mounted on the robot, can be seen in Figure CS 2.5. The two microphones are mounted on the top, facing forward, and their exact separation can be varied. In the experiments below it was set at 18mm, approximately a quarter of the wavelength of the 4.7kHz sound which is the carrier frequency of *Gryllus bimaculatus* song. The circuit is programmable so that both the relative delay between the ears and the gains of the combined signal can be altered. In the experiments below the relative delay was set to 53 microseconds – the time for sound to propagate a quarter of a wavelength at 4.7kHz – and gains of 1 were used. The electronics can potentially be expanded to include another two microphones representing the spiracle input, but these have not been used in the work described here.

Modelling the neural system

Crickets have 50–70 auditory receptors near each tympanum which vary in frequency and intensity range (Huber 1983). The axons form the auditory nerve, which runs from the foreleg tibia to the prothoracic ganglion. A number of pairs of auditory neurones have been identified in this ganglion (Figure CS 2.6). For simplicity I will here discuss only one pair (designated AN1 by Wohlers and Huber 1980) that have been critically associated with phonotaxis behaviour. Crickets appear to turn to the side of the more 'strongly' active AN1, and hyperpolarizing this AN1 reversibly alters the walking direction. The dendrites

Figure CS 2.5. *The robot used in the experiments.*

of this pair arborize with the axons of auditory cells tuned to the calling song frequency, and the tuning curve of the cell is centred on this frequency. The amplitude of a syllable of sound is coded both by the number of spikes fired and the latency to onset of the first spike, over a range of about 40 dB (Schildberger 1984). There is no background activity, and excitation causes a rise in potential with superimposed spikes; firing is maintained for the duration of the signal, and the recovery rate is relatively slow, estimated at about 15 ms to decay 50% (Wohlers and Huber 1980).

There are several features to note. Firstly, either the number of spikes or the relative latency of spikes could be used to determine which way the animal turns. The first AN1 to start firing will represent the side closest to the sound, as will the AN1 that fires most. Neither mechanism has been ruled out to date, although two problems with the latter mechanism are (1) crickets will apparently turn towards the signal occurring first if equal amplitude signals are presented to each side (Pollack and von Helversen, personal communication) and (2) crickets turn away from the side receiving a continuous signal towards a

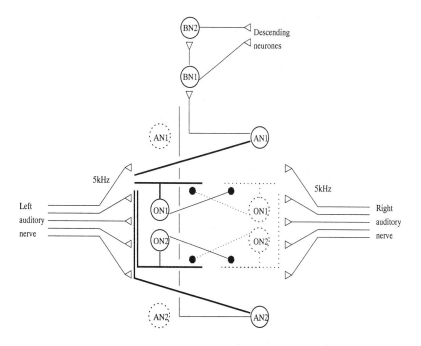

Figure CS 2.6. *Schematic description of known neural connections in the cricket auditory system.* All connections are bilaterally reflected but for clarity not all are shown.

patterned signal that produces a lower firing rate (Stabel *et al.* 1989). The latter result has led researchers to hypothesize that the AN1 signal turned to is the one that provides 'best input' to a recognizer (Pollack 1986). However, the simpler hypothesis that it turns to the first onset is also consistent with this evidence, as a continuous signal will not contain onsets.

Secondly, the slow recovery rate means that the firing pattern of these neurones cannot accurately encode rapid patterns in the signal. The cut-off point for accurate encoding appears to be in the vicinity of the rate at which signals are no longer tracked by the cricket. One possible interpretation is that the cricket no longer tracks the sound because it cannot recognize the signal in the continuous AN1 response. Another is that the use of first onset to decide which way to turn is no longer possible so the cricket may stop tracking because it cannot localize the sound rather than because it cannot recognize it. Firing rate, on the other hand, is not obviously affected by this issue: a continuous signal will still produce a firing rate difference so an additional mechanism is needed to explain why approach to the signal ceases. AN1 neurones send axons to the cricket brain where some further auditory neurones have been identified. Some of these have the interesting property of responding more strongly to certain syllable repetition rates — either high-pass, low-pass, or bandpass — and some of the latter have a response curve that apparently closely matches the probability of tracking in the cricket, leading to their interpretation as 'recognizers' (Figure CS 2.7) (Schildberger 1984). The actual mechanism of filtering is unknown: a possible interpretation that does not treat them as recognizers is given in the model below. Descending neurones involved in phonotaxis have also been identified but their anatomical and functional connectivity with the sensory neurones is not known. No central process for comparison has yet been identified.

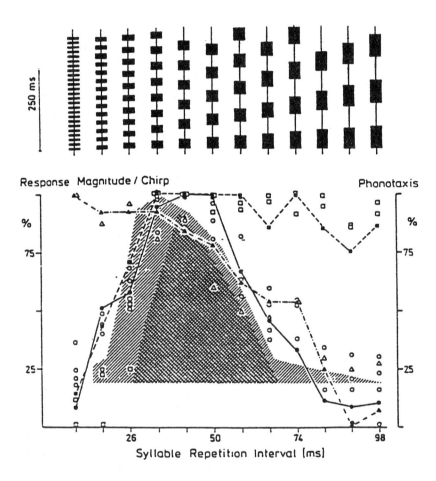

Figure CS 2.7. *Phonotaxis and neural responses as a function of syllable repetition rates.* The shaded areas show the strength of phonotaxis as a function of the syllable repetition interval. The lines show the recorded responses of neurones with high-pass, low-pass, and bandpass characteristics; the similarity of the bandpass responses to the strength of phonotaxis led to the relevant neurones being labelled as 'recognizers'.

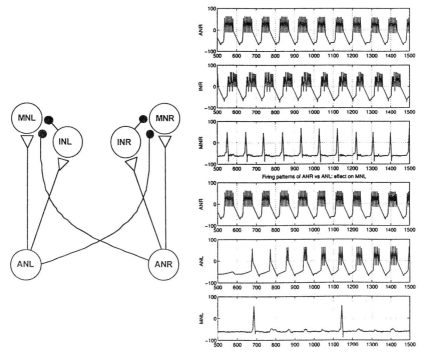

Figure CS 2.8. *A neural model for the control of phonotaxis.* The six-neurone network on the left was used to simulate a phonotaxis response. The traces on the right show the simulated firing patterns for a sound on the right. Each burst of spikes in ANR results in one spike in MNR, and inhibits the opposite MNL.

Hypothesis to be tested

A simple hypothesis for the control of phonotaxis is the following: the onset of AN1 firing on one side initiates a turn signal in that direction and suppresses the opposite side. Turn signals occur only at onsets, so continuous activity in AN1 caused by continuous or rapid syllables does not lead to approach behaviour. A certain frequency of turn signals are needed to cause an effective response, so slow syllables do not lead to approach behaviour. A neural model of this mechanism is illustrated in Figure CS 2.8. Whichever auditory neurone AN receives stronger input will fire first, causing a spike in the motor neurone MN, inhibiting the opposite MN and activating the interneurone IN such that further response from MN will be inhibited until AN has ceased firing. While only the AN neurone in this circuit is closely based on an identified cricket neurone, the MN neurone turns out to have properties not unlike the brain neurones described above, with the highest 'firing rate' to a bandpass of syllable rates. This is because rapid syllables lead to almost continuous firing in AN and so both MN are suppressed; and slow syllables produce less frequent onsets and so less frequent spikes in MN.

This 'neural circuit' has been explored in simulation using model spiking neurones, but to date is implemented on the robot in a simplified algorithmic form. In this, the two ear inputs are fed into leaky integration processes representing the auditory neurones, and the first to reach a threshold activates the respective motor neurone. The motor neurones are also represented as leaky integrators, so that repeated activations are needed to cause a turn.

The controller is obviously a simplification of the cricket system; the question is whether it is a valid simplification, i.e. correctly describes the basic functional mechanism behind phonotaxis behaviour, or whether it leaves out essential elements. In particular, is this simple structure, which lacks an explicit recognition process, able reliably to approach sounds and distinguish between correct and incorrect sounds in a manner resembling the cricket?

Experimental design

A critical reason for using a robot to explore this hypothesis was that it could be tested with real sound input. The aim was to carry out the following experiments based on cricket studies:

1. Sound localization. Firstly the robot must be able to localize sound reliably. This is a basic test of plausibility. The robot should be able to move towards the song of male crickets and do so reliably even in somewhat 'noisy' environments. (It should perhaps be noted here that not all female crickets 'reliably' track sound, although it is assumed the failure reflects motivational factors rather than disability.) By starting the robot from different positions in its arena (see below) relative to a speaker playing recorded cricket song we required it to turn and track towards the speaker. In experiments with the prototype robot, the basic hypothesis was shown to work but using a slowed down version of cricket song, with syllables lasting 300 ms rather than 30 ms. One reason for repeating these experiments was to answer criticisms that this slow signal made the task significantly different from the cricket's task. In particular, with a slowed version of the song, the robot could move too far between syllables. This meant it had to be able to respond to every syllable which crickets do not appear to do.

2. Preferred carrier frequency. A number of studies have shown frequency selectivity in cricket phonotaxis. Popov and Shuvalov (1977) reported that *G. bimaculatus* have a threshold 10 dB lower for 4.5 kHz song than for 3 kHz or 7 kHz and no taxis occurs above 12 kHz. Segejeva and Popov (1994) reported even sharper tuning: 15–20 dB threshold difference between 5 kHz and 6 kHz or 4 kHz song. Stout *et al.* (1983) reported similar sharp tuning to 4–5 kHz in

arena tests on *Acheta*, and Moiseff *et al.* (1978) used flight tests to demonstrate tuning for 5–5.5 kHz in *Teleogryllus*. Oldfield (1980) in a more detailed study testing not only thresholds but actual turns made reported that only chance behaviour is observed for 2.5 kHz or 12 kHz, and the angle from the midline at which significantly more correct turns are observed increases as frequency increases from the calling song. Thus at 4.5 kHz the cricket can turn correctly to song 13° from the midline, at 5.5 kHz it must be 30°, and at 6.5 kHz it must be at least 50°. This finding, along with the 'anomalous taxis' (tracking but at the incorrect angle) described by Thorson *et al.* (1982) suggests that the decrease in tracking to incorrect sound may be related to a decrease in localizability rather than recognizability. This can be tested by our robot as it does not have any direct filtering for song frequency.

3. Preference for syllable rate. Female cricket preference for a bandpass of syllable rates has been established in a number of experiments (for example Popov and Shuvalov 1977; Thorson *et al.* 1982; Stout and McGhee 1988). Thorson *et al.* (1982) argue for a '30 Hz' hypothesis (for *Gryllus campestris*), claiming that a syllable repetition rate of around 30 Hz is necessary and sufficient for taxis in the female, whatever the length of syllables or higher order grouping of syllables. They found that if the repetition interval between syllables was less than about 20 ms or more than about 50 ms, the female would no longer track the sound. Doherty (1985) and Wendler (1990) suggest that chirp structure alone may also be necessary or sufficient for taxis (at least for some female *G. bimaculatus*), but there is agreement that syllable rate is a primary cue.

4. Ability to choose between sounds. The data on female crickets' choice behaviour is less clear-cut than the preferences discussed above. What seems clear is that they can selectively approach one sound out of two in natural and arena situations. Weber and Thorson (1988) claimed that crickets on a treadmill also track directly to one sound out of two, randomly switching between them. If one sound is a few dB louder it will be approached. Wendler (1990) has argued that crickets will track between songs of equal amplitude, but the reported data — mean tracking angles — do not make this clear. Preference tests for songs with different frequencies or temporal parameters provide rather mixed results. Ryan and Keddy-Hector (1992) have argued that all demonstrated preferences can be subsumed under a preference for signals of higher energy content, implying that choice simply reflects the workings of the localization system. This is what we tested here.

The common paradigm adopted was based on the 'arena' experiments on crickets, where a cricket is released on a flat area with one or more speakers producing calling song, and the path it takes recorded. In our experiments here the arena was a 240x240 cm area of hardboard in the centre of the lab floor. No specific soundproofing was used. The speaker played either cricket song or computer-generated song with varied parameters as described below. The robot was usually started 1–2 metres away from the speaker, alternately from each side, facing 45° away from it (i.e. across the arena — see tracks below). Tracks were recorded with an overhead camera and a customized program that gave the floor co-ordinates of a bright light on the robot in real time.

Results

1. Localizing cricket calling song (Figure CS 2.9)

Male *G. bimaculatus* song was recorded and played back through a speaker. The robot was started five times from each side of the speaker and in every trial turned and tracked directly towards it. In a number of unrecorded tests the robot was able to move to the speaker from a variety of locations in the arena and even able to track towards live *G. bimaculatus*. This was a useful confirmation that it was not just the simpler slowed-down signal that had allowed the original robot to perform taxis but that the mechanism was adequate to control behaviour with real stimuli.

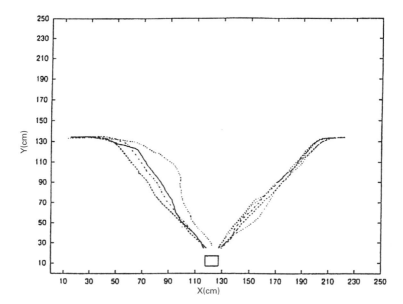

Figure CS 2.9. *The robot successfully recognizes real cricket songs.*

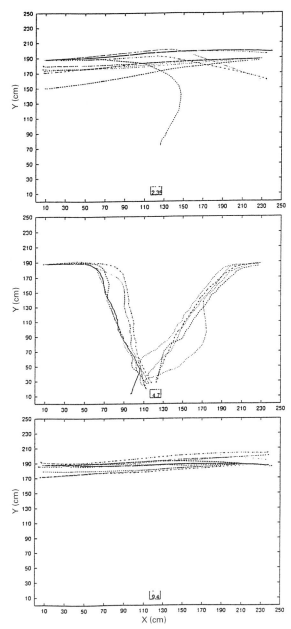

Figure CS 2.10. *The robot's response to songs of different carrier frequency.* Top left and top right: 2.35 kHz and 9.4 kHz. Bottom: 4.7 kHz.

2. Carrier frequency selectivity (Figure CS 2.10)

In the following tests we used synthesized cricket song. The temporal parameters were fixed with three 20 ms syllables at 20 ms intervals occurring every 200 ms. The carrier frequencies used were 2.35 kHz, 4.7 kHz, and 9.4 kHz. The auditory circuit was calibrated to remove any differences caused by the frequency response characteristics of the microphones. Two tests were done using different starting distances from the speaker. For each starting distance there were ten trials for each frequency, five from each side of the arena. Although the robot never turned to 9.4 kHz it could respond to 2.35 kHz – but only if it came within 1 m of the speaker. At 4.7 KHz, however, the robot could track the speaker from 2 m or more, suggesting the effective threshold (as amplitude of signal drops with distance) was lower. In other words the robot showed a comparable preference for the correct frequency to crickets although it had no pre-filtering for frequency; it simply could not extract directional information if the frequency was incorrect. How closely the 'preference' of the robot can be shown to match the cricket requires further tests – however, it is likely that the simplifications made in the current model (for example, using only two sound openings) may limit the ability to show quantitative as well as qualitative similarity in the behaviour.

3. Syllable rate selectivity (Figure CS 2.11)

In these tests we again used computer-generated song. We used continuous trills, that is, songs without chirp structure consisting of syllables repeated at different rates. The syllable periods used were 10 ms, 20 ms, 30 ms, 40 ms, 50 ms, 60 ms, 70 ms, and 80 ms, with a constant duty cycle of 50%. The robot was started about 1.5 m from the speaker, again alternating sides, in ten trials for each syllable period. The results show that the robot performed successful taxis only to syllable periods of 30 ms, 40 ms, and 50 ms. At 20 ms and 60 ms the robot would occasionally make some turns but only once did this suffice to reach the speaker. At 10 ms, 70 ms, and 80 ms there was no reaction to the sound source. In other words the model described is sufficient to cause the robot to have a bandpass preference for syllable rate. A similar result had been shown with the previous robot. Apart from using cricket-speed syllables, the main difference here was a sharper cut-off for the slower syllables, caused by the use of the MN process that prevented any reaction to single syllables.

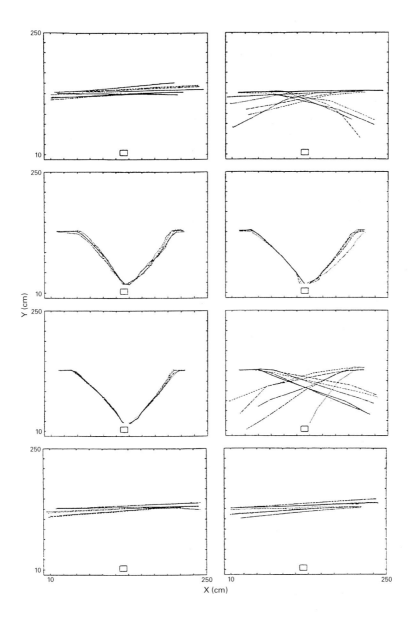

Figure CS 2.11. *Syllable rate selectivity.* The robot's behaviour in response to songs of different syllable rates. From left to right, and from top to bottom: syllable intervals of 10 ms, 20 ms, 30 ms, 40 ms, 50 ms, 60 ms, 70 ms, 80 ms.

4. Choosing between sounds (Figure CS 2.12)

One of the more interesting results with the previous prototype robot had been the finding that when presented with two sounds it was able to move directly to one of them, producing tracks that looked like 'choice' though clearly no real choice was being made. It appeared the interaction of a two-sound source field and the implemented mechanism was sufficient for one sound to capture the response. The experiment was repeated with the new robot. Two speakers were placed about 1.5 m apart and the robot started about 1.5 m equidistant from both and facing between them. When both speakers played the same 4.7 kHz synthesized cricket song the robot moved quite directly, four times to one and six times to the other speaker. However, when one of the songs was changed to 6.7 kHz then the robot always moved to the 4.7 kHz sound source. Further experiments to test whether the robot can choose between different repetition rates of song are planned.

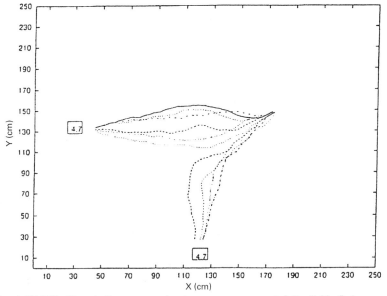

Figure CS 2.12. *The robot's response when two songs are presented.* It reliably finds one or other sound source.

Conclusions

Animal communication systems have provided a rich seam of research in ethology. They can be investigated on a number of levels, from evolutionary questions of mate choice and honest signalling to physical and physiological details of signal production and reception. They involve both basic sensory and motor processes, behaviours, and decisions. In some cases learning is involved, and they can be seen as the evolutionary forerunners of language. Consequently they make good model systems for the study of behaviour. One such communication system is the mating behaviour of crickets.

What are the main conclusions to be drawn from the results of the robot experiments? The methodology is rather different from the usual hypothesis-testing modes in biological investigation so it is necessary to be careful about what has been 'discovered' by this robot implementation. Ultimately, models are like theories in that, no matter how well they explain current data, they may still turn out to be false. Physical models are sophisticated means of developing hypotheses and deriving predictions from them: the ultimate arbiter is reality. What has been shown is the sufficiency of the simple hypothesis: using phase addition for directionality and first onset for turning control is sufficient to reproduce the basic forms of phonotactic selectivity seen in crickets. In other words, cricket phonotaxis may be simpler than was thought, in particular requiring no explicit recognition of the correctness or attractiveness of signals, as this is subsumed under the localization mechanism.

Further investigation of crickets suggested by this method include the following questions: (1) What relative role is played by the inherent frequency tuning of auditory neurones vs. the frequency limitations of directionality in the auditory system? (2) Are time differences in AN firing alone able to cause turns? For example if a short syllable on one side preceded a normal length syllable on the other, then latency and number of spikes would suggest opposite directions of movement. (3) Can recognition be disambiguated from localization in neural responses? There are a number of studies that have used stimuli that fail to allow this distinction to be drawn. (4) How should female choice studies be interpreted? The results here suggest that a simple sensory bias hypothesis may suffice to explain the fact that females can directly approach the sound more like the correct calling song. However, better experiments on both crickets and the robot are needed to follow this up.

The robot described is by far the most detailed and explicit theory of cricket behaviour that has been offered. There have been some mathematical models of parts of the system (for example Eilts-Grimm and Wiese 1984; Horseman and Huber 1994) but none have tackled the actual production of behaviour. It should be noted that it would be relatively easy to adapt the robot model to reflect alternative hypotheses provided these were sufficiently well specified. The work done here has highlighted the gap between 'understanding' a function and actually being able to implement it. The process of implementation, particularly in a physical system, was vital in suggesting mechanisms as much as testing them.

End of Case Study 2

Spatial orientation

A particularly interesting class of spatially based animal activity is that leading to the production of structures, especially nests of various kinds. Any structure, whether excavated or fabricated, requires that some parts should be built before other parts. For example, when building a house, the roof cannot be built until the walls that are to support it have been built. It is also necessary that the parts that are built early in the sequence should be shaped to accommodate the features that are to be added later, or the building procedures that are to be executed later. Construction of new structures by humans involves the preparation of blueprints, the detailed planning of an appropriate sequence of activities, and the correct performance of that sequence. Where there are several builders working on the same construction, their activities must be appropriately co-ordinated. All of this requires quite an abstract sense of space, and an understanding, at least at the planning stage, of how the part currently under construction relates to the future whole. However, when many identical buildings are to be constructed one after the other, it is easy to see that an alternative strategy will become possible – each building could be constructed through the repeated performance of some stereotyped sequence of activities, with no requirement for planning or spatial insight required on the part of the builders.

Obviously, the animal approach to building must be rather different, but some of the basic ideas can be transferred to the animal context. For example, it might be that something corresponding to a blueprint is inherited by an animal of a given species, and that it uses its intelligence and basic skills, perhaps in conjunction with learning, to produce something that corresponds to the blueprint. At the other end of the scale, an animal might inherit the stereotyped sequence of actions that, if carried out correctly, will usually produce a certain type of structure. The control of the sequence might be internal, with the completion of each action triggering the start of the next, or external, with the completion of each action producing some environmental change which in turn produces a sensory effect triggering the next action. In the case of some animal constructions — for example, birds' nests — it is possible to investigate the nature of the system controlling the building behaviour by interfering with the construction, either while it is in progress or after it has been completed, and observing the resultant behaviour. Such investigations have produced ambiguous results, in birds at least; no one strategy — blueprint or stereotyped action sequence — accounts for all observed results (Hansell 1984).

In the case of social insects, applying human ideas of organizing construction becomes much less plausible. Some termite mounds are enormous; some wasp nests are constructionally very complex indeed. It is clearly impossible that any single insect could even apprehend the whole structure, let alone compare it to some internal blueprint. On the other hand, how could such large, elaborate, and diversified structures be produced by simple stereotyped actions or chains of actions, especially when there is a need to co-ordinate the separate actions of the large numbers of builders working simultaneously? The answer to this question is somewhat surprising: through self-organization. The idea that systems can self-organize to produce patterns was first proposed to account for various physical and chemical phenomena in which patterns observed at a macroscopic level emerge from many local interactions between elements defined at a microscopic level (Nicolis and Prigogine 1977; Haken 1983). Recent work has shown that the theories developed to describe these processes can be extended to describe many of the characteristics seen in social insect colonies, with individual insects playing the parts of the microscopic elements, and spatiotemporal patterns in the environment constituting the macroscopic components requiring explanation (Bonabeau *et al.* 1999). Such diverse phenomena as nest construction in termites, raid patterns in army ants, and the foraging behaviour of bees all turn out to be manifestations of self-organization. As far the general field of spatially based behaviour is concerned, the message is very clear: spatially complex behavioural outcomes do not necessarily imply the involvement of equally complex processes of sensing or orientation. Our next case study illustrates this very neatly; it also demonstrates the successful use of a multiple robot system, a technique which may turn out to be of great value in studying social insects, and social behaviour in general.

Case study 3: Gathering and sorting in insects and robots *by Owen Holland*

Gathering and sorting in social insect colonies

The behaviour of social insect colonies is remarkable not only for its complexity and appropriateness, which often appears to imply the existence of some form of intelligence, but also for the way in which the actions of many individual insects — up to 20 million in the case of the army ant *Eciton burchelli* — are co-ordinated to yield a coherent outcome. Typical end results include the construction of nests of appropriate size, effective foraging, co-ordinated brood care, the management of elaborate cultivations, adequate cleaning and maintenance, and apparently appropriate group responses to minor and major disturbances (Franks 1989; Hölldobler and Wilson 1990; Robinson 1992). When we think of managing such enterprises at the human level, we think of strategies such as hierarchical organizations, teams, planning, explicit task allocation, task monitoring, and so on. However, it is clearly inappropriate to suppose that social insects may be using these methods, as they lack the necessary cognitive and communicative abilities. How, then, is the behaviour at the colony level organized and controlled?

The answer, which has taken shape over the last few decades, is that such behaviour is self-organized: the progressive and cumulative effects of simple interactions between individuals, and between individuals and the environment, yield behavioural complexity and co-ordination without any requirement for any other factors. The idea of self-organization was first developed within physics and chemistry to account for the emergence of macro-scale patterns from interactions between micro-scale components. Although individual social insects are orders of magnitude more complex than physical or chemical components, it turns out that much of their behaviour, like that of particles, can be described in terms of simple rules, often of a stochastic nature, relating local responses to local stimuli. (Such responses may include elements of short-term memory or learning.) The basic principles of self-organization are becoming increasingly well understood. Its key elements have been identified as positive feedback, negative feedback, amplification of random fluctuations, and multiple interactions; the characteristic signatures of its involvement in a process are the creation of spatiotemporal structures in homogeneous media, the possible coexistence of several stable states, and the existence of parametrically dependent bifurcations (Bonabeau *et al*. 1997, 1999).

Many species of social insects use self-organization to perform tasks that involve gathering objects together (clustering), and/or sorting different classes of objects. A typical example of a clustering task is the construction of cemeteries, or aggregations of ant corpses, by species such as *Lasius niger* (Chrétien 1996) and *Pheidole pallidula* (Deneubourg *et al.* 1991). (See Chrétien 1996 for a review.) If a colony is suddenly presented with a large number of corpses scattered throughout the environment, ants will begin to move the corpses, picking them up one at a time, carrying them for a while, and then dropping them. After a time, small heaps of corpses will appear in several locations; as time progresses, the number of heaps will get smaller, often reaching one, and the sizes of the heaps will increase correspondingly. The locations of the eventual large heaps are often related to certain environmental features (e.g. they are often found against walls) but the exact positions are typically unpredictable.

Sorting tasks usually centre around the sorting of the brood: after an experimental or natural disturbance, such as a nest migration, the eggs, microlarvae, larger larvae, prepupae, and pupae will be moved around within the new nest, and spatially distinct groupings of each category will emerge. Typical outcomes include sorting into heaps (Deneubourg *et al*. 1991) and sorting into concentric rings (Franks and Sendova-Franks 1992). The spatial density with which each category is packed may vary systematically with the category; in the case of concentric rings, the order in which the rings occur (measured from the centre) may be fixed.

A simulation model of clustering and sorting

A simulation model dealing with both clustering and sorting was proposed by Deneubourg et al. (1991). Each of a number of agents, which they pointed out could be regarded either as robot-like ants (RLAs) or ant-like robots (ALRs), performed a random walk over a square network of points. The network contained a number of objects, either of a single type, or of two types; at the beginning of each simulation run, the objects were scattered randomly across the network. Each ALR could pick up and carry a single object at a point, and could drop it at some other point; the probabilities of picking up or dropping an item were modulated by the type of object to be picked up or dropped, and the ALR's estimate of the local density of each type. (The estimate was based on temporal integration of information from objects directly encountered on the random walk – requiring a form of short-term memory – but an instantaneous estimate based on some analogue of olfaction or vision would give the same results.) The scheme for modulation of the probabilities was simple and elegant: as the estimated local density of a given type of object increased, the probability of picking up an object of that type decreased, and the probability of dropping such an object increased. The results showed a strong qualitative similarity to natural clustering and sorting, in both the changes in distribution of objects over time, and the final arrangement.

Although abstract gridworld simulations can be very useful, when successful, in demonstrating the operation of self-organizing processes, they do not necessarily make a strong case for the operation of those processes in corresponding situations in the real world. There are two main problems. The first is that the simplifications made in abstracting the model away from the real world may have omitted some factor present in the real world which might prevent the operation of that particular self-organizing process. For example, the real world is full of correlated noise; simulations almost always use uncorrelated noise. The second is that the characteristic simplified structures used in simulations (discrete space, discrete time, synchronous updating) may introduce artefacts peculiar to those structures. For example, many phenomena observed in synchronous cellular automata disappear in the closely related class of asynchronous cellular automata. For these and other reasons, if a self-organizing process hypothesized to occur in social insects can be shown to occur in a group of suitably programmed autonomous mobile robots, it carries a greater degree of conviction than a gridworld simulation ever could.

Robot clustering

The first successful robotic investigation of self-organized clustering was that of Beckers et al. (1994); a related subsequent study of interest is Maris and te Boekhorst (1996). Beckers et al. used between one and five small robots which

Figure CS 3.1. *A U-bot.* These 25 cm diameter autonomous mobile robots are fitted with infra-red object sensors (the three small boxes mounted facing straight ahead, and obliquely to the left and right), and with a bulldozer-like gripper for manipulating Frisbees.

could each pick up and move one or two small pucks around an arena, but which would drop the pucks and change direction when they detected that they were attempting to move three or more. This corresponds to detecting the local density as being less than three, or not less than three. The robots could also detect one another, and the arena walls, using infra-red proximity sensors; these triggered a turn away from the detected object. Between changes of direction the robots followed nominally straight courses. This simple set of behaviours was sufficient to support a self-organizing process that would gather the initially scattered 81 pucks first into many small piles, then into fewer larger piles, and finally into a single cluster, in a way qualitatively comparable to natural ants' performance with corpse collection. The process was robust: robots could be added or removed at any stage, and clusters could be disturbed by the experimenters, but the end result was inevitable. It is particularly interesting that this was achieved with a simple threshold estimate of local density, which contains less information than that used in the simulation of Deneubourg *et al.* (1991).

Robot sorting

This general approach has recently been extended to deal with self-organized sorting in a group of robots.

The robots (Figure CS 3.1), known as U-bots, are typical microprocessor-controlled autonomous robots of about 25 cm diameter, equipped with standard infra-red obstacle avoidance sensors set to detect obstacles at a range of about 20 cm and a height of around 10 cm. They operate in a large octagonal arena; a video camera mounted 6 m above the centre of the arena is linked to a PC-based video capture system which automatically records an image at preset intervals to

Figure CS 3.2. *A U-bot gripping a Frisbee.* If the U-bot turns, the shape of the gripper and the barbels (visible just inside the rim of the Frisbee) will retain the Frisbee during the turn. If the U-bot reverses, the Frisbee will either be left, or dragged backwards, depending on whether the Frisbee-retaining pin has been lowered or not.

facilitate analysis. Instead of pucks, the robots manipulate Frisbees, which are placed on the floor concave side up. The obstacle sensors are unable to detect the Frisbees.

A gripper mounted on the front of each robot enables it to sense, grip, retain, and release Frisbees. If a robot moving forwards encounters a Frisbee directly in its path, the Frisbee will fit neatly inside the semicircular part of the gripper (Figure CS 3.2), and will then be moved forward by the robot. A passive mechanism retains the Frisbee when the robot turns on the spot. When the Frisbee is right inside the gripper, an optical sensor detects its type. Two types are used: rings, which have either a white centre and black surround, or a black centre with white surround; and plains, which are either all black or all white.

The functionality of the gripper is extended by two further mechanisms. The first is a pin mounted at the rear of the gripper, which can be lowered by a small electric motor so that it projects down inside the concave rim of a Frisbee being pushed by the robot. If the pin is in the raised position, and the robot reverses, the Frisbee will be left in position; if the pin is lowered, and the robot reverses, the Frisbee will be pulled backwards. The second mechanism is similar to that used by Beckers *et al.* (1994): the gripper is suspended so that a backwards-acting force greater than some threshold value can move the whole gripper and trigger a microswitch. The force may be delivered when a Frisbee being pushed by the robot strikes another Frisbee or an obstacle, or when the jaws of an empty gripper strike some Frisbees or an obstacle. The threshold value is adjusted so that the microswitch is not triggered when the gripper strikes or pushes a single Frisbee, but is always triggered by two or more.

Self-organized clustering

The first task was to see whether the new robot and gripper arrangement, combined with a suitable algorithm, could produce the self-organized clustering observed by Beckers *et al.* (1994). The robots were programmed with a rule set similar to that used by Beckers, but with a slight difference imposed by the nature of the arena boundary. Beckers' arena had a deformable boundary wall which enabled the robots to push their grippers past pucks lodged on the boundary, and to scoop them up. The new boundary is made of rigid plastic pipe, and we found that it was extremely difficult to adjust the infra-red obstacle sensors to the exact distance necessary to enable Frisbees left on the boundary to be retrieved reliably: the robots would either turn away before having picked up the Frisbees, or would become stuck against the boundary, prevented by the Frisbee from approaching close enough to trigger the obstacle sensors. The modified rule set, in decreasing order of priority, became:

Rule 1:
if (gripper pressed & object ahead) then
    make random turn away from object

Rule 2:
if (gripper pressed & no object ahead) then
    reverse small distance
    make random turn left or right

Rule 3:
go forward

    The rules work as follows. A U-bot approaching another U-bot or the boundary will detect it, but will continue moving until it hits it with its gripper. (The U-bots are designed to withstand frequent collisions.) The combination of the gripper force rising above the threshold and an obstacle being detected will then trigger a random turn away from the side on which the obstacle was detected. If the U-bot is pushing a Frisbee, it will retain the Frisbee during the turn. If a U-bot's gripper is pressed in the absence of an obstacle, as happens when it attempts to push more than one Frisbee, the U-bot will reverse for a short distance (just over half a Frisbee radius) and make a random turn; the effect of this is to leave the Frisbee at the location where the gripper was triggered. If neither of these conditions is met, the U-bot moves forwards in a straight line.

    If a Frisbee is touching the boundary, a U-bot heading straight for it will receive it into its gripper, which will immediately be triggered; since the boundary wall will have been detected, the U-bot will obey Rule 1 and turn through a random angle away from the boundary, keeping the Frisbee in its gripper. It will then set off in a straight line (Rule 3) in the new direction, taking the Frisbee with it. The U-bots can thus remove single Frisbees from the boundary. However, if a Frisbee lies so that it is touching a Frisbee that is in turn touching the boundary, a U-bot that runs into the first Frisbee will have its gripper triggered, but will be too far away from the boundary to sense it. It will therefore obey Rule 2, and will back off the Frisbee, leaving it in place.

Forty-four Frisbees were placed in the arena at a uniform and regular spacing. (In anticipation of the sorting work reported later, half of the Frisbees were painted black (plains) and half were painted white with black centres (rings); at this stage, the robots were programmed to treat both types alike.) Ten robots were released, and the video system was set to record a frame every 5 minutes. The Frisbees behaved in the same way as Beckers' pucks, first aggregating in small clusters, then forming larger ones, and finally, after 8 hours 25 minutes, forming a cluster of size 40, which we had decided was an acceptable criterion for completion. (Beckers *et al.* [1994] used a criterion of 100% clustering; it was felt that this would unduly prolong the experiments, and around 90% would be a better cut-off point, given that the natural end point is typically a dynamic equilibrium.) Figure CS 3.3 shows the end state of an experiment that was allowed to run on until the Beckers *et al.* criterion was reached. It is not surprising that the system behaved in the same way as Beckers' system: the main difference is the threshold setting for triggering the gripper (two Frisbees rather than three pucks) and the relative sizes of the Frisbees (23 cm) and pucks (4 cm).

Figure CS 3.3. *The end state of a clustering experiment.* This is a view taken from the ceiling-mounted camera. Nine of the ten robots used in this experiment are in the upper half of the picture.

However, we noticed that in the middle phase of the experiment the robots had great difficulty 'stealing' Frisbees from the intermediate-sized clusters. The problem seemed to be geometric in origin: even in a small cluster, unless the form of the cluster was extremely ragged, it was rare for a robot running on a straight line trajectory to strike the cluster at a point that would allow a Frisbee to be removed but would not trigger the gripper.

The pull-back algorithm

Following on from the observation that the slow progress in the clustering task might be due to the excessive stability of intermediate-sized clusters, we decided to modify the algorithm so that Frisbees would not always be deposited hard up against one another, producing dense clusters, but would be spaced apart. To do this, it was necessary to employ the pin-dropping mechanism to enable robots to pull the Frisbees backwards for some distance before releasing them. To avoid producing clusters with too diffuse a structure, we decided to use both ring and plain Frisbees, but to apply the pull-back tactic only to the plain Frisbees.

The rule set now becomes more complex:

```
Rule 1:
if (gripper pressed & object ahead) then
    make random turn away from object

Rule 2:
if (gripper pressed & no object ahead) then
    if plain then
        lower pin and reverse for pull-back distance
        raise pin
    endif
    reverse small distance
    make random turn left or right

Rule 3:
go forward
```

The U-bot's behaviour with respect to rings is unchanged. However, if it is pushing a plain and hits another Frisbee, or if it collides with a plain Frisbee that is already on a cluster (perhaps at the boundary), the new version of rule 2 will cause it to drag the Frisbee backwards and leave it-the pull-back distance away from the contact point.

We used 22 ring and 22 plain Frisbees, again uniformly spaced throughout the environment at the start of the trial; six robots were used, with the pull-back distance being set at 2.6 Frisbee diameters.

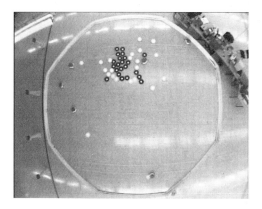

Figure CS 3.4. *The first appearance of sorting using the pull-back algorithm.* After 7 hours 35 minutes, 90% of the ring Frisbees are in a single cluster, in a visibly more compact arrangement than the plain Frisbees.

Figure CS 3.4 shows the distribution of Frisbees after 7 hours 35 minutes, the earliest time at which 20 of the 22 ring Frisbees were in a single cluster. (We defined a cluster as a group of Frisbees, of any combination of colours, in which every member was within a Frisbee radius of at least one other member.) There is a central dense core of 16 rings, with 11 plains and the other four rings being packed around this core, and the other plains scattered more loosely nearby. This is clearly an outcome that can be regarded as sorting of some kind. We had conceived the algorithm as a means of improving the rate of clustering, and had not thought that it might achieve sorting on its own. However, since the algorithm was probably simpler than anything we might have devised, we decided to explore it further. This serendipity is characteristic of experimental work using robots, especially when robots are repeatedly interacting with the products of their own behaviour: it is extremely difficult to predict the outcomes of experiments, and the outcomes, often highly structured and consistent, may undergo radical changes following apparently inconsequential or irrelevant changes to the robots' sensors, programs, or environment.

The experiment was allowed to run on for half an hour; the cluster became more compact, but the number of rings and plains in the cluster at 8 hours 5 minutes was unchanged. The progress of the experiment over time reflected the final state in miniature, with a small number of tight ring clusters each surrounded by plains gradually giving way to the eventual pattern. Figure CS 3.5 shows the distribution of Frisbees after 1 hour 45 minutes.

86 Artificial ethology

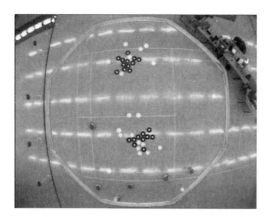

Figure CS 3.5. *Intermediate sorting produced by the pull-back algorithm.* After 1 hour 45 minutes, two partially sorted clusters have been formed; these are eventually merged into the single sorted cluster of Figure CS 3.4.

To check that this was not a fluke, we repeated the experiment four times. We defined the end of a trial as the time of occurrence of the first cluster containing at least 20 rings, and measured the elapsed time, and the number of plains included in the cluster. In every case, the outcome was similar, in that a single cluster containing 20 rings was formed. However, in the replications, the distribution of plains within the cluster was generally less good, in that some plains were trapped within the body of rings.

Trial	1	2	3	4	5
Time in hours	7.58	2.75	25.3	11.7	4.50
Number of plains	11	12	11	10	12

Table 1. Data from five trials of the pull-back algorithm, showing the times to completion, and the number of plains included in the terminal cluster.

Table 1 shows the times to completion (when 20 rings were in the same cluster) of all five trials, and also the number of plains included in the cluster. The number of plains is almost constant; however, the times are strikingly variable, with the slowest time (25 hours 20 minutes) being more than nine times the fastest (2 hours 45 minutes). Examination of the video records revealed that the very long times were associated with the formation of two clusters of the same size at opposite sides of the arena; these clusters could remain quite stable for hours.

Varying the pull-back distance

The obvious next step was to vary the pull-back distance to establish any effects on the quality of clustering, the time to completion, and the quality of sorting.

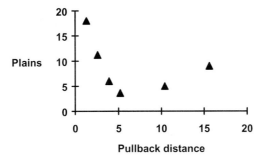

Figure CS 3.6. *The average number of plain Frisbees included in the final cluster as a function of the pull-back distance.*

Single trials were run using distances of 1.3, 3.9, 5.2, 10.4, 15.6, and 26.6 Frisbee diameters. In every case a single cluster of rings was formed, except for the trial using 26.6 diameters, which appeared to be making no progress whatsoever after 6 hours, and was terminated. The number of plains in the single cluster varied systematically with pull-back distance, decreasing from 18 (at 1.3 diameters) to a minimum of three (at 5.2 diameters) and increasing again to ten (at 15.6 diameters). We added a further four trials at the 'best' value (5.2 diameters). Figure CS 3.6 shows the (average) number of plains in the final cluster for each trial.

The paucity of data points at some values makes interpretation difficult, but it may be reasonable to summarize the data as follows:
- for pull-back distances of 15.6 or less, a single cluster containing at least 20 ring Frisbees is eventually formed;
- the number of plain Frisbees forming part of the cluster decreases with increasing pull-back distance to a minimum near 5.2, and then increases;
- for a given pull-back distance, the time to completion can vary considerably.

A key additional factor is the spatial distribution of the plains that are not members of the central cluster. (Let us call them detached plains.) From observation three things are clear:
- detached plains tend to be found near the final cluster;
- as the pull-back distance increases from 2.6, the detached plains tend to become more widespread in the arena. (This is difficult to judge for values up to 10.4, simply because there are more detached plains.);
- detached plains are often found near the two detached rings at termination.

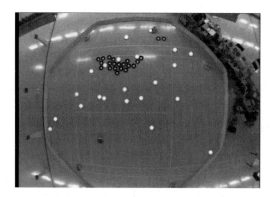

Figure CS 3.7. *Good segregation at a pull-back distance of 5.2 Frisbee diameters.*

How should we judge the ability of the simple pull-back algorithm to produce sorting and segregation? It is clear that the algorithm can produce good segregation when the pull-back distance is around 5.2 Frisbee diameters (Figure CS 3.7). It is also clear that a degree of annular sorting is also occurring in some trials, with rings being concentrated in the centre of the cluster, and plains being found at and beyond the edges of the cluster. However, the values giving good segregation are not the values giving the annular sorting.

At this stage several questions arose. Would plain Frisbees show any degree of spatial association if there were no ring clusters to act as foci? Could not segregation alone be achieved easily and economically simply by ignoring the plain Frisbees, and using the basic clustering algorithm on the rings? And what would happen if, instead of determining the pull-back distance by the colour of Frisbee, it was determined randomly? All these were tested in a series of experiments.

Will plains cluster in the absence of rings?

Twenty-two plain Frisbees were uniformly spaced throughout the arena, and six robots with the pull-back distance set to 2.6 Frisbee diameters were released. After 6 hours there was no sign of any static cluster whatsoever, and the experiment was terminated. However, for much of the time most of the Frisbees tended to be loosely grouped together, with most of them within an area of perhaps 30% or 40% of the arena. Such an arrangement would dissolve on a time scale of some tens of minutes, only to reform in another location. To check that continuing the experiment would not have led eventually to a tight cluster of plains, we performed an additional experiment, using the same robots and Frisbees, but starting the Frisbees off in a tight central cluster. Within a short time, the central cluster had been broken up; the experiment was continued for 6 hours, and for the remainder of the time it behaved exactly like the first experiment.

Since the pull-back distance used was relatively short, and had previously produced some of the more convincing sorting outcomes, we concluded that:
- a static focus of ring Frisbees is necessary to anchor spatially the dynamic cluster of plain Frisbees;
- for pull-back distances greater than 2.6, the arena may be too small to enable any plain clustering to be seen.

Can segregation be achieved by ignoring the plains completely?

Within the confines of our paradigm, the plain Frisbees cannot be ignored passively, because Frisbees are acquired by the robots in a passive way. Instead, a plain Frisbee that has been collected in the gripper must be released immediately. This is achieved by modifying the rule set — paradoxically, by making it more complex.

```
Rule 1:
if (gripper pressed & object ahead) then
    make random turn away from object

Rule 2:
if (plain OR [gripper pressed & no object ahead])
then
        reverse small distance
        make random turn left or right

Rule 3:
        go forward
```

This does not mean that plain Frisbees will never be moved. They can be moved for short distances in several ways: when being pushed forwards by the outside edge of a gripper before either drifting outside the gripper's jaws or entering the gripper; by being pushed forwards during the brief interval between the Frisbee colour being detected and the motors coming to a complete halt; by being pushed backwards by a robot executing Rule 2 after encountering some other Frisbee; by being pushed by a robot executing Rule 1; and by being pushed by another Frisbee which is itself being moved. Almost all of these already occur in all the experiments described, merely forming part of the noisy context of any physically instantiated system.

Once again, 22 Frisbees of each colour were uniformly spaced throughout the arena, and six robots were released. To begin with, matters seemed to be progressing as in most of the other experiments, but after a time it became clear that the plain Frisbees, which were being moved around a small distance at a time by effectively random contacts, could not easily be removed from the boundary once they had arrived there. In turn, ring Frisbees dropped next to these plains also proved difficult to remove. This experiment was terminated after 18 hours; no single cluster was ever formed, but mixed aggregations of plains and rings constantly formed and dispersed close to the periphery.

Applying the pull-back algorithm randomly to rings and plains

To see the effects of applying the pull-back strategy randomly to both rings and plains rather than always to plains alone, we modified the robot program to determine the variable 'Frisbee colour' at random when the gripper was triggered, rather than by sensing. The starting arrangement was as in Experiment 3, with 22 Frisbees of each colour spaced uniformly over the arena; six robots were used, with the pull-back distance set to 5.6 Frisbee diameters. This experiment corresponds exactly to what we had first thought of as the solution to the cluster stability problem; however, our experience with the previous experiments now led us to expect that a single tight cluster would not be formed, and that instead a rather looser and mobile assembly would be seen. We were again mistaken: the termination criterion of 40 Frisbees in a single tight cluster was reached after 15 hours 15 minutes. We were so surprised by this result that we ran an additional experiment using the 'ring' strategy alone, again using 44 Frisbees. This reached the termination criterion in 8 hours 25 minutes — less than twice as fast as the mixed strategy, although this may not be a representative observation, given the variability seen in these experiments.

On close examination of the recordings, the reason for obtaining a single tight cluster became clear. It is simply a function of the relative ease of leaving a Frisbee on the cluster, compared to taking one off. A Frisbee left hard against a cluster by the 'ring' behaviour has a relatively small probability of being removed by a robot under Rule 3, as has been noted previously, but may be removed to the pull-back distance by a 'plain' behaviour under Rule 2. A Frisbee left at the pull-back distance from a cluster by the 'plain' behaviour is equally likely subsequently to be pushed onto the cluster by a robot using the 'ring' behaviour, or pushed onto the cluster and then pulled back by a 'plain' behaviour. Since the probability of a 'ring' or 'plain' behaviour is equal at 0.5, all that is necessary to achieve tight clustering is that the probability of being removed from a cluster by a 'plain' behaviour must be less than the probability of being added to a cluster by a 'ring' behaviour. This is clearly the case, and so tight clustering is to be expected.

Discussion

The simulations of Deneubourg *et al.* (1991), directly inspired by observations of ant behaviour, showed that sorting could be achieved by a system of simple homogeneous agents with the ability to sense the local density of each type of object to be sorted. The work reported here, growing directly out of the experimental manipulation of a system of homogeneous mobile robots, shows that a form of sorting can also be achieved when the agents have no ability to estimate the local density of each type of object. Three questions arise: How does the sorting work? Are there any indications that such a method might be employed by social insects? And what conclusions can we draw about the utility of using collective robotics to investigate problems of social insect behaviour?

The way the sorting works is extremely simple. There are two components: clustering, and sorting within clusters. The mechanism of clustering is identical to that operating in the Beckers *et al.* (1994) experiments: the larger the cluster, the more likely it is to acquire Frisbees (pucks) removed from other clusters, and the less likely it is to lose Frisbees (pucks) itself. To understand sorting, consider a robot that is approaching a cluster, pushing a Frisbee. The robot must approach the cluster from the outside, moving in a straight line, and there must therefore be a positive resolved component towards the centre of the cluster. Regardless of the colour of the Frisbee, it will penetrate the cluster until the gripper is triggered. A ring Frisbee will be left at that point; since the reversing robot will now have a negative resolved component towards the centre, a plain will be pulled towards the outside of the cluster before being dropped. Now consider a robot without a Frisbee approaching the cluster. If it encounters a loose Frisbee, ring or plain, it will scoop it up and continue to move towards the centre of the cluster; when it collides with another Frisbee, it will leave a ring Frisbee at the site of the collision, but will leave a plain further away from the centre. Ring and plain Frisbees therefore penetrate equally far into clusters, but the plains are subsequently moved towards the outside of the cluster with which they are associated, whereas the rings tend to remain in situ. The cumulative effect of the operation of these tendencies produces the sorting in both intermediate and final stages.

Could the results obtained here have any significance for the understanding of ant behaviour? There have been no reports of this strategy having been observed in use, but the elements of the technique used here can certainly be found in the literature. For example, an account of *Leptothorax* building behaviour (Franks *et al.* 1992) mentions the possible use of an increased resistance to pushing a building block forwards against other building blocks as the cue to drop it: '...if the ants drop their granule only if they meet sufficient resistance...'. Franks *et al.* also report a behaviour that is reminiscent of the pull-back algorithm: '...workers individually carry granules into the nest. They walk head first towards the cluster of their nestmates, who are already installed in the nest, forming a fairly tight group. After coming close to the group of ants, the builder then turns through 180° to face outwards from the nest. The worker then actively pushes the granule it is carrying into other granules already in the nest or, after a short time, if no other granules are encountered it simply drops its load.'

As far as the potential utility of using collective robotics for the investigation of ant behaviour is concerned, it should be clear that both the work of Beckers *et al.* (1994), and the work carried out with the system described here (Melhuish *et al.* 1998; Holland and Melhuish 1999) have demonstrated methods for producing self-organized clustering and sorting that are novel, robust, and compatible with the known abilities of ants, yet are simpler than the strategies discovered by direct observation of ants, and subsequent simulation. At the same time, the demonstration of these strategies in a real physical system gives assurance that they are robust in the face of real-world noise. These considerations support the idea that the use of collective robotics in the study of social insect behaviour is a valuable and valid addition to existing techniques.

Acknowledgments

Much of the research and experimental work described here was carried out by Chris Melhuish. Ian Horsfield built the U-bots, and he and Jason Welsby kept them running for thousands of robot-hours.

End of Case Study 3

The three case studies in this chapter have shown robots to be valuable tools for studying sensory processes and orientation, at least in arthropods. This is perhaps not surprising; the essential quality of a robot is to be able to move around in response to sensory input, and the exploration of the consequences of simple sensor-based control of movement is a natural first step. In Frank Grasso's study, the robot also offered a powerful method of investigating a challengingly complex environment from the animal's point of view. All three studies indicate new directions for experiment and observation on animals which would be most unlikely to have been discovered without the use of robots.

Chapter 4

Motor co-ordination

The nervous system is responsible for controlling behaviour and, to a certain extent, for controlling the animal's internal environment. This control is exercised by commands to muscles and glands.

Muscles are made up of complex protein molecules that are capable of contraction and relaxation. Nerve endings connect with muscles by means of synapses similar to those by which neurones connect with each other. The nerve impulses arriving at the neuromuscular junction set up electrical potentials, which cause the muscle to contract. Muscular relaxation results from lack of stimulation. When the muscle contracts it becomes shorter, provided it is not prevented from doing so by being fixed at each end. A muscle can lengthen

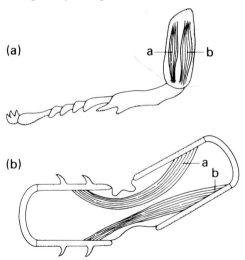

Figure 4.1. *Mechanical arrangement of muscle and skeleton in an insect's leg.* The muscles are housed inside the skeleton. In (a), muscle a is the extensor (bending the limb) and muscle b is the extensor (straightening the limb). In (b), where the muscles span the joint, the arrangement is the opposite. Muscle a is the extensor, and muscle b is the flexor.

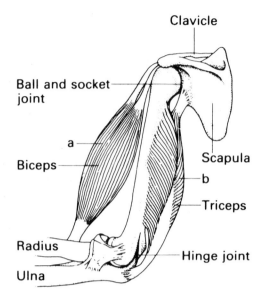

Figure 4.2. *Mechanical arrangement of muscle and skeleton in a human arm.* The muscles are arranged outside the skeleton. Muscle a is the flexor and muscle b the extensor.

when it is relaxed, but only if it is stretched by the action of other muscles or by some extraneous force. Thus muscles are usually arranged in opposing groups which act against each other. In some invertebrates, such as annelid worms, muscular contraction may be resisted by the hydrostatic pressure caused when the muscles squeeze part of the body cavity. This pressure causes the muscles to lengthen when they relax. In other invertebrates, such as arthropods, the muscles are housed inside a hard exoskeleton, which provides the necessary leverage for opposing sets of muscles (see Figure 4.1). In vertebrates the internal skeleton provides the leverage and the muscles are arranged so that they pull against each other (see Figure 4.2). One set of muscles relaxes when the other contracts.

The somesthetic system

It is important for an animal's brain to receive information about the state of the body. The positions of limbs, the pressure on internal organs, the temperature of various parts of the body, and many other factors are monitored by the central nervous system via interoreceptors located at strategic points. This system, responsible for bodily sensations, is called the somesthetic system.

Numerous types of sensory receptors exist in the skin, skeletal muscles, and viscera of vertebrates. Invertebrates also have a wide range of receptors. Humans have five types of skin receptors, giving rise to sensations of touch, pressure, cold, heat, and pain. The pain receptors are numerous, being 27 times as common as cold receptors and 270 times as common as heat receptors. Some skin receptors show rapid sensory adaptation. In response to a step change in stimulation, the frequency of nerve impulses rises rapidly and then declines to its resting level. This means that the receptor is a good indicator of changes in intensity of stimulation but a poor indicator of the absolute level of intensity. This is an advantage in cases where the skin receptors give early warning of environmental changes that are likely to affect the body, like changes in temperature.

Receptors deep inside the body serve a wide variety of functions including detection of changes in blood pressure, the tension of muscles, the amount of salt in the blood, etc. We humans are not directly aware of the information produced by the majority of interoreceptors. They do not give rise to sensations.

Sometimes their effects combine to produce sensations of hunger, thirst, or nausea, but these are due to complex processes in the brain, which do not always refer the sensation to particular parts of the body. This is presumably because the action that has to be taken in response to hunger and thirst is much more indirect than action that is taken in response to peripheral touch or temperature change.

The orientation of animals in relation to gravity and external stimuli like light depends partly upon information about the spatial relationship of the various parts of the body. In mammals this information comes from the vestibular

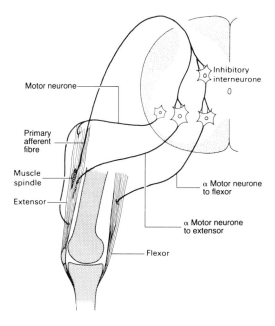

Figure 4.3. *Mammalian muscle spindle.*

system and from receptors in the joints, muscles, and tendons. The joint receptors provide information about the angular position of each joint (Howard and Templeton 1966). In the tendons of mammals are the Golgi tendon organ receptors, which are sensitive to tension. They send messages to the spinal cord and are involved in a simple reflex that acts to oppose increases in muscle tension.

Within the muscles are the muscle spindles that are sensitive to changes in muscle length (Figure 4.3). The muscle spindles consist of modified muscle fibres that have a spiral nerve ending, called the primary (or annulospiral) ending, wrapped around their middle. When the muscle increases in length, the muscle spindle is stretched and the primary endings send fast messages to the spinal cord. There may also be secondary flower-spray endings that send slower messages. Many mammalian spindles have both primary and secondary endings, while others have only primary endings (Prosser 1973). These spindles form part of a simple reflex that acts to oppose increases in muscle length.

The muscle spindles are contained within a fusiform connective tissue, and their muscle fibres are called intrafusal fibres. Ordinary muscle fibres are called extrafusal fibres. These are innervated by alpha motor neurones, which have their cell bodies in the spinal cord. In mammals the intrafusal fibres are innervated by smaller gamma motor neurones, which keep the spindle in a tonic state of activity so that less muscle stretch is required to activate the spindle. Since muscle spindles are arranged in parallel with the extrafusal muscle fibres, they tend to become slack when the muscle contracts. The gamma motor neurones can command the intrafusal fibres to take up the slack so that the spindle remains in a state of readiness (Figure 4.4).

The muscle spindles of birds resemble those of mammals, with the intrafusal and extrafusal fibres in parallel. In the lizard *Tiliqua*, however, the muscle spindles appear to be in series with the extrafusal fibres (Prosser 1973). In reptiles and amphibians there is no gamma motor neurone system, and both intrafusal and extrafusal fibres are supplied by alpha motor neurones. Fish have

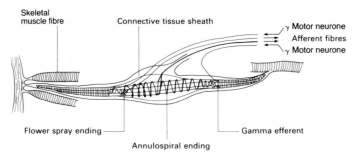

Figure 4.4. *The innervation of the muscle spindle.*

no muscle spindles, but they do have receptors in their fibres that are sensitive to their angular velocity.

In the next case study we see how an apparently simple amphibian reflex movement is controlled by a highly complex system, and how a robot has played a part in uncovering the mechanisms involved.

Case study 4: How frogs groom   by Simon Giszter

Frog wiping behaviour

Frog wiping behaviour has been utilized in the study of motor control since the 19th century. The frog's wiping responses are largely organized in the spinal cord: they persist generally unaltered in a spinal frog, i.e. after spinal transection. Wiping behaviours are attractive for research because the goals of the behaviour are well defined, the motions are repeatable, and the computations and transformations required are a simple subset of those involved in, for example, visually guided behaviour.

Remarkably, many of the frog's grooming and protective reflexes can be produced by the spinal cord alone. Frog wiping behaviour can be simply elicited in the spinal frog by a small piece of paper soaked in acid or a local electrical stimulus. The frog removes the irritant with a graceful co-ordinated motion of the hindlimb involving some seven degrees of freedom.

In different areas of skin several qualitatively different reach strategies are necessary for removal of the stimulus. In these cases the frog chooses a strategy adaptively. Some of these skin areas overlap, so that one of several possible strategies must be chosen. The choice in these regions is based on the overall configuration of stimuli on the body and the history of responses executed prior to the current response. These areas in which a configuration can be chosen from one of several disjoint sets of joint angles have been called transition zones.

A second adaptation which occurs for multiple stimuli on the skin is blending of responses. In blending responses the frog adapts the kinematics of two different movements elicited in different areas of skin in order to reach both stimuli in a single movement.

Finally, within a strategy the frog can make small adjustments of configuration, presumably in order to optimize the relationship of the wiping effector to the irritant. These adjustments are continuous across an area of skin.

The circuitry that is used to perform these sets of movements is fully contained in the spinal cord. This has made the spinal frog an attractive preparation with which to examine mechanisms of limb positioning and control. Understanding the neural architecture that supports this set of behaviours may help in the design of artificial creatures and mechanisms.

Frog wiping experiments have been broadly organized around three conceptual themes: (1) understanding the biological implementation of ill-posed sensorimotor transformations (ill-posed problems are those for which many different solutions are possible); (2) understanding the modularity of vertebrate movement and its organization; and (3) understanding the control strategies organizing limb motions. Each of these broad themes is best understood in relation to wiping by examining the whole array of wiping behaviours and their organization.

Following the Marr (1982) scheme of analysis of a task, a behaviour can be examined as a task specification, as the requisite computations for the task, and as a specific implementation. In very general terms the task of wiping consists of locating a stimulus on the body surface, and executing an action that removes it. Generally the animal subdivides the task: it first closes the kinematic chain consisting of the body segment with the target stimulus and the effector, and then executes a movement that removes the irritant. Computational aspects of this task thus require the frog to transform skin location, limb configuration, and body scheme so as firstly to select an appropriate effector, secondly select an associated set of postures and limb trajectories for the effector, and then thirdly generate a set of appropriate muscle activations to control and move the effector through these postures and trajectories to the stimulus and remove it. All of these stages can involve ill-posed problems, and so one solution must be selected from the many that are possible.

The frog's body surface can be divided into a set of regions within which different wiping strategies are used. A strategy as defined here is a choice of an effector, a final posture, and a pattern of kinematic folding of the limb. Some of these regions overlap. In the region of overlap, transition zones exist in which one of two strategies can be chosen. In such a transition zone both strategies are competent to remove the irritant. This represents motor-equivalence. This phenomenon was first described in the turtle (Stein *et al.* 1986) and replicated in the frog (Berkinblit *et al.* 1989). It is not known whether the borders of transition zones are defined by the movement limitations of joints or by the extents of the skin's sensory fields. These might of course coincide. Switching or blending can occur for multiple stimuli in different zones. Responses in transition zones depend both on the history and the context of stimulation.

Wiping strategies can be divided into responses in which a limb wipes a body surface, and responses in which the limb is wiped on the substrate. An adhesive or persistent irritant can be first transferred onto the surface of the foot or leg itself from the body surface. The irritant can then eventually be transferred onto the ground by a substrate wipe. When a limb is wiping a body surface, the effector region may consist of the entire hind foot from the heel down, the distal joints alone, or an intermediate area. Similarly the number of distinct kinematic phases and the degree of involvement of joints in wiping motions varies depending on skin region and effector. Experimentally, the wipes that have been

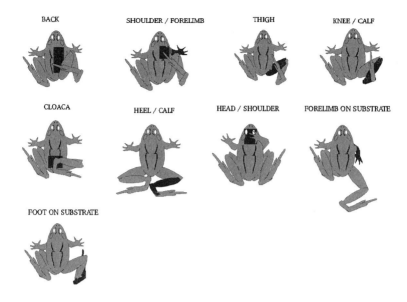

Figure CS 4.1. *Different skin zones elicit different wipe patterns.* This shows a partial summary of the zones eliciting qualitatively different wipe types, and the effectors used for those zones. The first six all involve the hind-limb; the head wipe uses the forelimb; the last two wipes involve rubbing the limb on the substrate.

best examined are those to the back, forelimb, hind-limb, and cloaca. Some typical wipes are shown in Figure CS 4.1.

Biological experiments

Fookson *et al.* (1980) found that in wiping to the forelimb, the spinal animal adjusted the responses of the hind-limb in accordance with forelimb position. It now seems likely the frog may have assumed the observed variety of forelimb postures actively (Giszter *et al.* 1989) but the data of Fookson nonetheless clearly show that body scheme information was used in the spinal cord. Each wipe type comprises a different set of sensory-motor integration and co-ordination problems. A different number of intervening links and different effects are used in different wipes. Some wiping transformations are well-posed and some are not. Many require the use of body scheme information.

Berkinblit *et al.* (1986) first pointed out the ill-posed problems of wiping in the frog and the possible motor-equivalence that might result. In wiping to the back, it was observed that the animal adjusts limb position continuously in response to stimulus position, and that the choice of limb configuration in this task was ill-posed. Detailed analyses of wiping to the back in frogs, both before and after spinal transection, showed that individual frogs exhibited fixed solutions to the problem of choosing joint angles in order to position the limb in the placing phase (Giszter *et al.* 1989). These fixed solutions generally involved independent linear relations between the joint angles in the effector limb and the rostro-caudal position of the stimulus. Many other strategies of angle choice were available, as was confirmed by Monte Carlo simulation. Rather than choosing a solution based on using the full information on stimulus location, the frog used only rostro-caudal location. While each frog's strategy was fixed and linear, the parameterization of the strategy used could vary from frog to frog.

While this body of work shows the spinal cord alone can solve ill-posed and non-linear problems in the context of back wiping, the result does not imply a general ability to solve these problems. Work on the hind-limb to hind-limb wipe by Giszter *et al.* (1989) using Watsmart motion recording showed that the spinal frog was able to solve the problem of hind-limb to hind-limb wiping for the whole workspace only in a restricted stimulus zone. The restricted zone allowed a simple fixed strategy roughly equivalent to matching joint angles between the target and effector limbs. Hind-limb to hind-limb wiping throughout the workspace would necessarily involve a non-linear transformation of configuration information between the target and effector limbs. This was apparently beyond the abilities of the spinal frog, but not those of intact frogs.

Bizzi, Mussa-Ivaldi, and Giszter developed a novel description of limb responses by collecting a force field from the limb. In essence, this approach associates a force vector generated under isometric conditions with a position in the frog's limb workspace. Forces were examined both prior to and during an attempted spinal behaviour, or as a result of some stimulation in the central nervous system. By obtaining a set of such records across the workspace a force field description may be constructed; see Bizzi *et al.* (1991) for a brief review. These measured force fields can be decomposed into a resting force field that represents the resting postural state of the leg, and an active force field representing the control field or forces generated as a result of the active limb response. The force fields examined in this way have several interesting properties which suggests a consideration of them as primitive control entities (see Giszter *et al.* 1993). Firstly, the active force patterns simply scale through time, so that the field at any point in time can be viewed as a scaled version of the field at other times. Secondly, the fields elicited by stimulation of the skin, and by stimulation in the grey matter of the spinal cord, correspond very closely. Thirdly, the total force field of the limb in the isometric case converges on a posture. This posture would be the final posture of the limb were it free to move in wiping. That this is indeed the case has been confirmed experimentally. The force field represents an attractor in configuration space which is apparently simply turned on and scaled up in strength (see Figure CS 4.2). In many schemes of motor control, this type of force field time course would not be observed. For example, to generate a torque pattern in order to produce a trajectory in the presence of different passive mechanical environments, different patterns and time courses of field structure would be expected. The

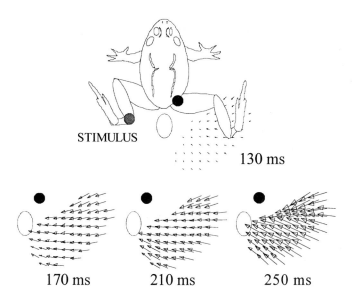

Figure CS 4.2. *Force fields supporting initial limb motion in a hind-limb to hind-limb wipe.* The filled circle indicates the hip, and the open oval is the region in which limb contact is achieved. The force fields are similar in structure over about 100 ms. The stability of structure is similar to the 'force field primitives' elicited by spinal cord microstimulation, and the patterns are also similar to one of the small set of primitives. The full trajectory involves a sequenced combination of fields.

implication of the behaviour of the force fields in the frog is that trajectory will not be conserved in different passive environments.

Recently we have shown that many postural phases and trajectory fragments of spinal behaviours can also be simply expressed as force fields (Giszter *et al.* 1991, 1993).

The force fields associated with three different tasks executed by a hind-limb of the spinal frog have been examined in detail. The force fields associated with these behaviours had fixed characteristics. They were hypothesized to represent movement primitives, by which we mean the smallest fragments of meaningful behaviours. We have also found that even spinal circuitry microstimulation produces fields that are remarkably similar to the fields underlying behaviour that are generated by skin stimulation. In summary, the main finding described here was that invariant force field patterns underlay each of these behaviours.

The spinally generated force fields were considered to include both position and force control strategies. The spinal cord thus appears to support both equilibrium point control strategies specifying absolute location, and other force field strategies specifying relative movements or contact forces. This suggests that there may be no explicit trajectory planning in the spinal cord: the path emerges from the interaction of limb dynamics with a force field. This force field results from environmental constraints, and also from the neurally generated force field. It has now been shown that in several instances these fields may exhibit vector summation when activated in parallel (Bizzi *et al.* 1991).

It seems that reflex spinal behaviours of some kinematic complexity might be constructed using simple fixed force field strategies. These force field or movement primitives might be thought of as permanently instantiated schemas for motor behaviour (see Arbib and Cobas 1991).

Synthesis of fields

One manner in which force fields might be combined is vector summation. In this mechanism of combination, the force at each point in the workspace following combination would be the vector sum of the forces of the individual fields. In principle this could be hard to achieve, given the limb and muscle non-linearities. However, a body of evidence indicates the use of this mechanism for field combinations in the frog spinal cord (see Bizzi *et al.* 1991).

The force field data discussed here allows some comment on the interpretation of kinematics. Even a framework consisting of fixed structure force fields predicts variable kinematics to a given target posture. The actual trajectory will depend on the initial conditions, the strength of the force field, and the overlap of the duration of force field activation with the activation of other force fields. The actual kinematics are described by a flow-field resulting from the interaction of the limb's inertia and its state, with gravity, limb viscosity, and the evolution of the actively generated force field. The relation of the force field to a given kinematic trajectory is thus also an ill-posed problem. The result of a chain of briefly activated and partially overlapping force field modules could thus easily result in neither the virtual trajectory nor the actual trajectory of the limb ever reaching any equilibrium posture in the sequence except for the final one. Thus kinematics alone is unlikely to be a good means of revealing or rejecting force field modularity in wiping.

How the spinal motor system is organized to provide the basic functionality described above in the spinal frog, and how the spinal elements combine to support the full behaviour, has now been examined in both theoretical and experimental studies. Not surprisingly, full attention to the physical aspects of the problem has proven critical.

Theoretical and simulation approaches

Theoretical work has attempted to address several of the issues raised above. I first review path planning and inverse kinematics models and then discuss models addressing issues of modularity.

One type of study has concentrated on the issues of motor-equivalence and solution of the ill-posed problems created by limb redundancy. The goal of these models has been to examine wiping in the broader context of the computational issues involved in general limb positioning. With regard to general limb positioning they succeed well, although it is likely the wiping mechanisms are not general purpose positioning mechanisms. Work in this area is well summarized by Shadmehr (1991) who has also provided a model. The goal of these models is to specify algorithms operating locally at the joint or muscle level that can solve ill-posed problems of kinematic planning.

The existing models assume a body-centred co-ordinate representation is available to the frog and proceed from there to the issue of obtaining a pattern of joint motions of a serially redundant linkage in order to lead the endpoint trajectory to a target. They are thus suitable for wiping to the back (where a body-centred co-ordinate representation is presumably directly available from skin location). Current models do not address how the initial body-centred stimulus representation is derived for stimuli on limbs in various configurations. Arguably, this is only an issue of reference frame, however. Both the Berkinblit and Shadmehr models use some minimization or endpoint compliance criterion to provide a particular choice of solution for the path-planning problem. The Berkinblit (1986) model uses gradient descent based on a simple independent rule for the motion of each joint. In the Berkinblit algorithm stiffness is constant in joint co-ordinates. The goal of the Berkinblit algorithm was to replicate the within-strategy motor-equivalence which they believed to exist. The model had the attractive feature of finding its way to the target location in the event of a joint constraint.

The Mussa-Ivaldi algorithm was used by Shadmehr. The particular value of this algorithm is that it generates inverse kinematics solutions that guarantee integrability across the workspace. At each point in Cartesian space the limb uses a single unique configuration or posture regardless of the route taken. This formulation thus can exhibit absolutely no within-strategy motor-equivalence once an initial configuration and endpoint stiffness have been chosen. Shadmehr chose straight line endpoint paths to the target stimulus. Using this algorithm to drive the endpoint motion to the target location, Shadmehr replicated the phenomenon seen in transition zones. He observed variable choice of behaviour strategy in the transition zone (i.e. different configuration or folding of the limb). In his model this variation in final folding is based on the practice of beginning trajectories from different positions with different motions of the limb. These different initial conditions at a point in the workspace lead to different solutions at the target location. The model as specified by Shadmehr in effect leads to the limb utilizing different solution manifolds for each different initial configuration. While the limb is operating on each solution manifold the chosen behaviour remains fully integrable.

Unfortunately neither the Shadmehr model nor Berkinblit's model is able to predict how the limb chooses trajectories in response to the effects of multiple stimuli nor how the scaled force fields observed experimentally are derived. It is not clear how to extend them in order to do this. These two models utilized the implicit notion of a single global mechanism for wiping and general trajectory planning. What type of force fields predictions the two models would generate is unclear.

A second class of models of frog movement deal with the use of the idea of force field primitives. Two classes of models have been developed based on functional and behavioural decomposition. Use of functional or basis fields, and vector summation will be examined first. These are relevant both to wiping and to the adjustment of wiping by brainstem and other descending pathways. Mussa-Ivaldi (1992) showed that arbitrary fields were readily approximated by a linear combination of conservative and circulating radial vector field primitives. The coefficients of these field primitives could be found by least squares or minimum norm methods.

Mussa-Ivaldi and Giszter (1992) applied the Mussa-Ivaldi vector field primitive work to the motor system in the non-redundant limb. Since this framework allows approximation of arbitrary smooth vector fields it follows that the planning and choice of the control fields to be approximated must be constructed in some other manner. Extensions of this static framework into the time domain are under investigation. Mussa-Ivaldi and Gandolfo (1993) used the basis field approach coupled to a neural network system to simulate use of functional primitives by a planner. One concern with this modelling direction was that the non-linear combination of endpoint force fields in serially redundant manipulators might pose difficulty for the use of this class of models in many biological systems.

Gandolfo and Mussa-Ivaldi (1993) tested how far the mechanism of basis field summation might apply to serially redundant linkages. They tested vector summation of endpoint force fields in a three-link planar serially redundant manipulator with elastic actuators. The manipulator thus had one excess degree of freedom in configuration space. The configuration and actuation spaces of the system were explored by using Monte Carlo methods. The results of this study suggested that in serially redundant linkages, near vector summation may occur among a large fraction of randomly chosen control primitives. They conclude that by selection from this subset the vector summation mechanism can be utilized. This analysis has yet to be extended to a serially redundant manipulator operating in three-dimensional spaces as is found in frog wiping behaviours.

I have explored the modelling of a system using behavioural primitives. The models form a counterpoint to functional approaches in that planning, decision making, and control of execution are all resident in and generated by a single network. Thus some degree of local autonomy and hierarchical possibility exists. Nonetheless, these models do not deal explicitly with redundancy issues. A serially redundant limb is either viewed as a single behavioural entity or as two connected and individually non-redundant behavioural entities. The degrees of freedom problem here is thus essentially a problem of selection and co-

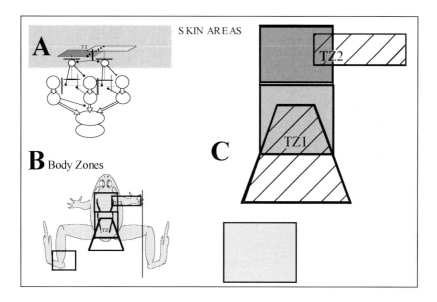

Figure CS 4.3. *Model skin zones.* A. Stimuli in model's skin zones act as input to a Maes net which sequences and combines primitives. B. Schematic of frog and relationship of chosen model skin zones to actual areas on a frog. C. Skin zones and areas of potential transition zones

ordination of non-redundant primitives. In the simulation a set of behavioural force field primitives was used to drive the virtual trajectory of a limb in order to simulate wiping to various regions of the body surface. The primitives were chosen and parameterized by an augmented Maes network (Maes 1989). This simulation raised a number of specific questions for biological experiments and the force field approach.

Modelling force field responses

If one hypothesizes that force field primitives can be taken as behavioural movement primitives then the critical issue for the nervous system is to choose sequences and combinations of these primitives that are both relevant to the situation and successful in satisfying the behavioural goals. The combination of force fields by vector summation appears to be an interesting mechanism to generate adaptive responses, but some summations will not attain the behavioural goal of either of the force field behaviours. It thus appears that some method of choosing allowable combinations of force fields is needed.

The task of the driving network for these primitives is then to select, sequence, and activate sets of primitives while preserving their behavioural relevance. Examining such models of the generation of wiping behaviours can fulfil several purposes: the model may suggest new experiments; the experimental data can form a critical test of model function; and models constructed during this theory/data dialogue that are not accurate models of the biological system may nonetheless be useful to artificial system development. One iteration of the data/model interaction is discussed below.

The model structure described here consists of three parts. The first is a simulation of areas of skin and associated sensory systems (Figure CS 4.3). These areas have associated overlapping sensory fields which transform skin stimulation into the activation of a layer of nodes which act as the conditions or propositions for input to the second element of the model, an augmented Maes network. The Maes network chooses a behaviour or combination of behaviours, which are implemented by the third component of the model, force field summation.

The choice of a Maes network for the core of the simulation was based on several criteria. The Maes network in its original form chooses a course or sequence of actions based on the global situation while using local interactions. Thus adaptive behaviour is an emergent property of the interaction of network and environment. In addition, there is potentially a simple relationship between the network structure, its global parameters, and neural entities such as pattern generators and modulators in the spinal cord. Finally, the one to one correspondence of nodes and actions seemed suitable for the activation of force field primitives.

The behaviour network design used here is built upon the organization originally proposed by Maes 1989 (and see Maes 1991a, b). It is worthwhile summarizing the core operation of the rather elegant basic network first.

The overall flow of activation propagated by the links leads to increasing activation of those behaviour nodes that are currently executable and also increasing activation of those nodes that will most likely shortly become executable as a result of environmental variations or current actions. Chains of nodes are therefore readied in anticipation of action. Thus threads and trees of actions and potentially useful actions emerge from the interactions of the network with the current conditions.

The augmentations added to the Maes network in the simulations here are described in detail in Giszter (1993). They consist of adding extended durations for behaviours, allowing concurrent operation of behaviours, introducing intensity variation of behaviours, adding an exclusion layer to the basic network, and allowing the network links to vary continuously rather than simply implementing on-off switching. The first three are essentially changes in the network output. The last two alter the structure and operation of the network.

The exclusion system controls access to the effectors. This mechanism prevents conflicting behaviours while allowing activation buildup to proceed unaffected. Concurrent execution is often desirable. Concurrency can be disastrous for some combinations of behaviours however. In the frog, two rostral and caudal wipe types cannot be combined and still retain any functional value. The exclusion network in this simulation is very simple. It excludes behaviours that conflict with a currently executing behaviour. Exclusion occurs

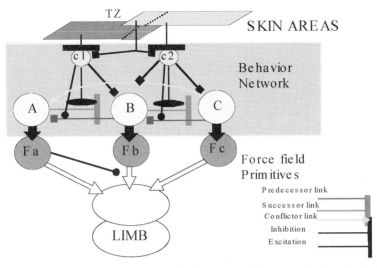

Figure CS 4.4. *The Maes behaviour network.* The Maes model is expressed schematically. The drawing convention attempts to link the network to a plausible biological circuit. Inhibition of weights involved in positive feedback between behaviour nodes controlling primitives is caused by sensory condition cells. Nodes A, B, C are directly activated by conditions. Conflictor and exclusion links inhibit or deny nodes access to the limb. Limb behaviour is driven by linear superposition of force field primitives activated by behaviour nodes.

on a first come first served basis. There is no pre-empting although this can be achieved very simply. The exclusion network design is handcrafted to avoid starvation and deadlock problems. However, the principal goal of the implementation used was simply to ensure that access to the effector system was separated from the process of activation buildup from which 'planning' emerges.

It would be possible to achieve exclusion with 'virtual' conditions. The reason for choosing explicit exclusion over the addition of novel 'virtual' conditions not anchored to the external environment is that this would destroy the biological analogy of conditions to sensory afferents, and more importantly that such 'virtual' conditions would alter the network operation substantially for behaviours that were mutually exclusive. Thus for example adding a 'virtual' condition to a flexion behaviour could have several results. Flexion normally follows extension. Thus it is a natural candidate for receipt of activation from extension but clearly it should also be excluded by extension, as their goals conflict. One could use a virtual condition to suppress the flexion. This would not allow flexion ever to become executable during periods of extension. However, flexion must become executable to promote activation of any third behaviour. For example, suppose the next behaviour were a locomotory push: its activation would be independent of the excluded flexion since the flexion would remain non-executable so long as it was excluded. The implementation here uses a separate exclusion system that allows such interactions. Thus more stable behaviour chains and limit cycles are anticipated in this network design. The exclusions proposed here were implemented as a simple message passing between the behaviour nodes. This implementation thus preserves the encapsulation of behaviours in nodes.

Virtual conditions in the type of network described here would allow hierarchical structuring and the subsumption of networks of this type into larger structures. This could be achieved by adding descending control of the executable/non-executable switch. Such a mechanism might be useful in the broader context of descending control of spinal motor systems (see below).

The Maes network augmented in this way was used to drive force field summation which formed the third and final element of the model. The limb's overall force field was used to calculate motions of a three-link limb. This limb was given the ability to remove all or part of irritant stimuli 'applied' to the model by moving the last link (the foot) across them in a defined direction. The effectiveness of the 'wipe' was proportional to the velocity of the motion. The overall simulation structure is summarized in Figure CS 4.4. Figure CS 4.5 shows the type of force field primitives used.

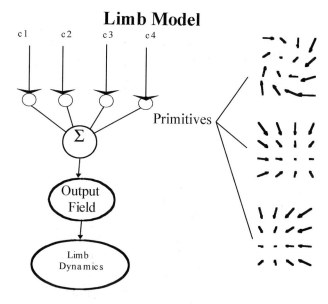

Figure CS 4.5. *Model of frog limb.* Primitives $c1$–$c4$ (examples on right) sum their force fields to generate an output field driving limb dynamics.

Simulation results

Although the network and limb were set up to deal with three-dimensional data, the principal results of this simulation are well expressed in two dimensions.

The simulation showed a very 'lifelike' set of response patterns. The main results demonstrated were as follows: successful stimulus removal; switching between multiple stimuli; choice of the strongest irritant on first wipe; blending of responses for multiple stimuli while preserving function; trial to trial kinematic variability of response; and apparent searching and exploration to reach 'difficult' stimuli. This latter exploration comprised a variety of activation patterns which caused the limb to cover more workspace and use trajectories not expected in the initial hand-coded design.

Stimulus removal for a single stimulus

The limb was moved close to a posture specified by a placing field and then it swept the area of the stimulus. For weak stimuli a small and ineffectual intention-like movement could precede a fully fledged response. Similar twitching and abbreviated response patterns for weak stimuli are observed in real frogs. After completion of the irritant removal, the simulation recovered the

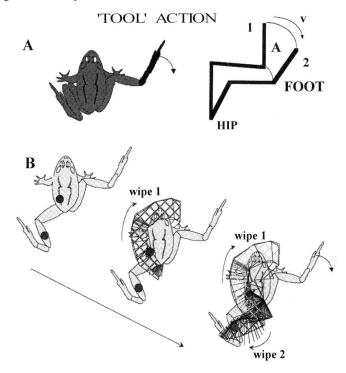

Figure CS 4.6. *Effector kinematics and sample model wipes.* A. The effector is the distal foot. This must hit the stimulus with the ventral surface to be effective (see arrow). Effectiveness (degree of stimulus reduction) is proportional to contact velocity. B. Sample wipes to two stimuli. The frog model switches between wipes 1 and 2. Where allowed, blending may also occur. V is velocity and A is angle

base posture and became quiescent for weak and moderate stimuli. For strong responses a short period of twitching and brief extensions were produced as the activation of the network died down. This type of behaviour is also seen in real spinal frogs and was called 'irradiation' by Sherrington. Figure CS 4.6A shows sample kinematics of the model.

Stimulus removal for multiple stimuli

Blending: When multiple stimuli could be reached in a single wipe using non-exclusive fields, these were utilized by the network. The network could also operate on a single stimulus initially and ignore the second. This depended on stimulus strength and prior network state.

Switching: Switching occurred when multiple stimuli could only be reached in different ways. This was not surprising since the exclusion mechanism explicitly coded into the network would be expected to generate this switching behaviour. Figure CS 4.6B shows switching and variable trajectories for strong stimuli.

'Exploration'

Difficult to reach regions were regions for which very precise and strong activation of single force fields or combinations were needed to bring the limb to the stimulus. These difficult to reach regions were due to my very approximate hand coding of force field positions. Interesting behaviour was generated in the simulation in these regions. The persistent flow of activation into the network from the almost unreachable stimulus coupled with a relatively sparse set of conditions and exclusions allowed a collection of increasingly variable responses in the vicinity of the stimulus to be generated. Whether to call this exploration or flailing is a moot point. This is true for both the simulation and for the real spinal frog which also exhibits similar increasingly violent responses. However, the net effect of these violent behaviours was usually stimulus removal. Presumably, a learning mechanism added to a network of this type could utilize the system behaving in this way. The learning system could learn the appropriate activation patterns for these 'hard' stimuli, and tune the behaviour network to cope with these stimuli more gracefully. This would form an interesting extension of this framework.

Learning

Problems of training the 'frog' system outlined above are numerous. The most severe of these are:
- several concurrent actions may be needed for success;

- success may require a sequence of actions of which only the last is rewarded;
- several stimuli interact in concurrent applications, so successful sequences and network dynamics depend on both stimuli;
- modification of the dynamics can be achieved in several ways.

If each action can be assessed for success, at each instance, the learning scheme of Maes (1991b) is appropriate. However, the work here describes a network embedded in a second layer of dynamics, owing to the vector primitives through which it acts. This additional layer does not allow a simple assessment of performance.

The learning scheme

Following execution of a successful action and the consequent reception of reinforcement, we should like the system to change as follows: decrease the time to execution of successful actions; and increase the domain of action-module state space in which the correct action execution is reached. There are difficulties in this approach, however. Ideally we need to do this with minimal interference in the effectiveness of the other sets of linear systems specified by the network that also participated in sequences that led to correct actions. Additionally, executing an action too soon may fail to achieve the goal in a complex dynamic environment. Presumably, however, there may often be local regions of the action-module state space in which performance can be improved.

The goal of the scheme chosen was to distribute small changes through all the relevant connections and in proportion to the connections' importance in the activation of executing modules, but decremented with distance from the executing modules. This may minimize the disturbance to other behaviours by localizing increasing change only in relevant links. The remainder of the network is very slightly degraded by the same reinforcement. Each link strength is slightly decreased. It is thus important that all behaviours occur sufficiently often on the average to maintain their performance. There is continual plasticity and stabilization of behaviour by their execution.

Relation of reactive and autonomous approaches to other frameworks

Current work supports the idea that primitives in the spinal cord (as described above) either specify position or force controls, but not trajectory details. Evidence from some experiments and the simulations presented here supports the idea that this description extends to wiping behaviours. These mechanisms involving local decision making by spinal circuits both constrain and simplify some aspects of descending control.

Other frameworks for use of force fields have been examined and are very powerful when provided with an explicit planning system (see Mussa-Ivaldi and Giszter 1992). The basis field approach described by Mussa-Ivaldi and Giszter allows generation of arbitrarily complex control fields by summation of a small set of primitives using linear summation mechanisms. This framework is ideally suited for implementing a descending control of spinal cord function in which all intelligent computation is performed elsewhere. The approach relies on vector summation of primitives in arbitrary combinations.

Since descending controls and autonomously organized local spinal behaviours must be seamlessly integrated into an adaptive whole, the rules of interaction between descending and local control may be complex. If the same primitives can be used in local behaviours and subsumed into a descending control scheme, the local behaviours must react appropriately to the context provided by the descending control and to the loss of those modules and actions that are co-opted by the descending pathways. Similarly the local behaviours must not prevent access to primitives needed by descending control, and must gracefully surrender these so as not to jeopardize the functioning of the system. The scheme presented here, based on the Maes approach, may allow the integration of local and descending control in a very flexible fashion. I would suggest that setting up the decision making and summation rules for the local reactive spinal circuitry in preparation for, and in support of, descending controls is an important aspect of normal motor control. Learning how to do this delegation may be a critical aspect of learning generally.

Summary and issues with the simulation

This simulation draws on recent discoveries in the frog spinal cord and has suggested several experiments. The simulation is able to deal with multiple stimuli, choose responses, and show a type of exploratory response. As discussed elsewhere, this network structure has much to recommend it as a plausible framework for modelling the spinal cord. In particular the global variables available in the behaviour network may have some correspondence to spinal observables such as sensitization, depression, habituation, and actions of neuromodulatory substances. The responses of the network are quite 'life-like' and fulfil the goal of providing an impetus and direction to experiment. It is unlikely that all the elements necessary to simulate spinal behaviours are present in the simulation but it encompasses a breadth of response patterns even with a relatively simple structure.

The model structure chosen in this last section consisted of three parts: the simulation of areas of skin and associated sensory systems, the augmented Maes network, and the force field summation. This type of processing maps fairly simply onto our ideas about possible sensory processing, and the regulation of the activation dynamics of the Maes net by presynaptic inhibition of internal links is not a great stretch.

The Maes network itself deserves closer scrutiny. In its original form it chooses a course or sequence of actions based on the global situation while using local interactions. Thus adaptive behaviour is an emergent property of the interaction of network and environment. There is potentially a simple relationship between the network structure, its global parameters, and neural entities such as pattern generators and modulators in the spinal cord. The one-to-one correspondence of nodes and actions seemed suitable for the activation of behavioural force field primitives. This is attractive. What biological structures and tests are implied?

The behaviour network is comprised of a collection of nodes. These represent behaviours or 'competencies'. Biologically these nodes might be thought of as pattern generators. The dynamics were simple, but the case for the inclusion of more dynamics based on appropriate cellular properties could be argued. It is not clear that the significant behavioural details would alter, however. Perhaps more germane is the behavioural sequences generated and the implications for organization of primitives. Individual primitives executed for fixed times. Thus timing dynamics were tied to each force field primitive in the simulation but phase relations among these were variable. This scheme is clearly testable and inspired experiments described below.

The choice of sequences of force fields and the strength of activation of the force field sequences was an emergent property determined by the network state and the sensory context. The model replicated blending, switching, and transition zone phenomena, and successfully dealt with multiple stimuli. Random exploration of allowed force field combinations was exhibited by the network in response to nearly unreachable stimuli.

The following questions are raised by the model:
- Are phase relations of biological force field primitives variable during wiping?
- Is a specific dynamics attached to a force field primitive in the real animal?
- Are the sequencing network and the primitive different structures or intimately connected?
- Is the fixed force pattern primitive sufficient to control the biological system?
- Is the biological spinal cord system plastic and does learning occur?
- How critical are details of temporal dynamics?

Robot experiments

To examine some of these questions we incorporated a robot able to mimic many aspects of frog limb motion into our experiments, using the robot to 'arm

Figure CS 4.7. *The small robot used to interact with the frog's limbs.* The robot is gripping a pen held by the experimenter. As well as exerting a programmable force on the gripped object, the robot records position data at 150 Hz.

wrestle' a spinal frog. We focused on the timing and stability of force field structure.

Several factors must be considered in trajectory formation. Conceivably, initial position and motion-related afference might modify the time course and duration of muscle activation and force development in the normal reflex trajectory, in support of the reflex goal. For example, in most goal-directed trajectory formation Fitt's law applies, relating movement duration, movement speed, and accuracy (Mottet *et al.* 1994). If the spinal frog's wiping trajectories obeyed Fitt's law with an accuracy constraint then the simpler force field pictures of trajectory formation would not be seen. This is because Fitt's law predicts that for movements to a single target that are of similar accuracy, the larger movements will have both higher velocities and longer durations than smaller movements (Mottet *et al.* 1994). It follows that the timing of force development would almost certainly differ during larger and small movements in a frog, even though this might not be detected in forces observed in isometric experiments. In our first set of experiments we therefore tested how Fitt's law applied to the spinal frog and whether there was a coherence of the timing of limb trajectories and muscle activity patterns among different unimpeded reflex movements. If afference during movement altered timing and pattern of force development in reflex movements with well-defined goals, these could not be described using force field primitives in a simple way.

Throughout all the experiments described here we used a small light-weight three degree of freedom robot (Figure CS 4.7) to interact with the frog hindlimb (Giszter *et al.* 1995). For trajectory timing experiments we positioned a spinal frog's ankle at a chosen location using the robot. We then programmed the robot to allow the frog's limb to move away from the holding locations with minimal impedance, as an electrical stimulus was applied to the frog's skin. The robot sampled the limb position at 150 Hz. The stimulus we applied was placed so as to induce wiping movements. The wiping movements we elicited always terminated in a fixed target zone caudal to the cloaca. Thus all the movements we examined had a common target but originated at different distances from this target.

We tested trajectories across a range of positions spanning the workspace (Figure 4.8). In this way we collected data describing limb trajectories originating from different distances, and three-dimensional locations relative to the target. We collected 7–15 trajectories originating at each location and compared these among locations using a two sample t test. We found that most features of the trajectories were similar regardless of location. The scaled tangential velocity profiles, the time of peak velocity, the time of peak muscle activation, and the total duration of the initial wiping movements were all similarly distributed at each location with similar means and standard

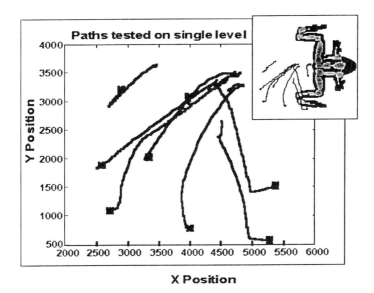

Figure CS 4.8. *Trajectories of paths tested using the robot.* The graph shows trajectories from seven points at a single level; the inset shows the trajectories in relation to the frog's body.

116 Artificial ethology

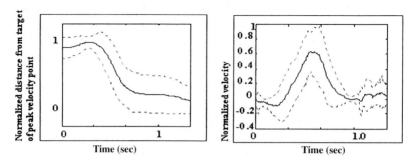

Figure CS 4.9. *Isochronous scaling in spinal frog trajectories from different areas of the workspace.* The fact that isochronous scaling occurs in the spinal frog enables the preparation of average normalized trajectory displacement (left) and velocity profiles (right).

deviations. None of these was significantly different from holding site to holding site in individual frogs (Figure CS 4.9).

Our results therefore differed from those predicted by Fitt's law with an accuracy constraint, but resembled results in reaching trajectories of cats in which it is known propriospinal interneurones may play a significant part (Altermark *et al.* 1993; Martin *et al.* 1995), or in collision-bounded voluntary human trajectories (Pfan *et al.* 1998). The reflex movements exhibited a strongly constrained time course: the spinal frog generated controlled trajectories of uniform duration across the entire reachable space. These data were reminiscent of the fixed timing of peak force observed in isometric force field measurements.

We next examined the effects of altering the motion-related afference by manipulating the limb's environment. The feedback provided to the frog's spinal cord by the unimpeded limb is plainly different from that generated in isometric experiments. Similarly, the muscles operate in a different region of their force-velocity-length surfaces when free motion is compared to isometric conditions. At the least, the force patterns elicited during movement should differ in magnitude from those observed under isometric conditions. Nonetheless, it was still possible that the force pattern structure that was observed during isometric conditions (the isometric force field primitive) was conserved during movement.

The controlled interactions of proprioceptive feedback effects with velocity-dependent muscle properties and limb motion could simply scale force fields observed in isometric testing. We sought to test this idea directly. Our approach was to use the three degree of freedom robot to present an external impedance to the limb during the behaviour. The work reported here is confined to the effects of linear elastic fields. Following stimulation and reflex activation the frog interacted with the robot-generated environment.

In effect the spinal frog and robot arm-wrestled as the frog attempted to execute a reflex behaviour. We found that in such experiments the limb always came to an active (isometric) equilibrium with the robot after a period of motion. At this point we measured the interaction forces between the limb and robot. Published work supports the idea that during the gradual deceleration towards the active equilibrium the active muscles return to isometric or near isometric conditions (Iwamoto *et al.* 1990). At equilibrium we therefore assumed that isometric conditions were established and that the force we measured closely reflected muscle neural activation. This experimental design enabled us to make a set of force measurements that were directly comparable to earlier isometric force field data. Note, however, that the measurements we made here differed from all previous data on spinally organized force fields in that they occurred immediately after a period of interaction with a controlled environment.

Potentially, timing variations could be a part of feedback-based interaction or compensation. To compare the forces generated by frog spinal reflexes in different environments it was therefore essential to establish that, just as for free limb motion and isometric force evolution, timing parameters did not depend on either the location or the different environments. We tested whether the robot-imposed environment perturbed the timing of the trajectory and underlying neural activity. We examined the timing of the active equilibrium between the frog and robot, and the timing of peak muscle activation in the externally applied fields as the limb moved from different starting locations. Once again, the timing parameters did not differ significantly, even in different environments. Furthermore, the timing of peak muscle activity was similar in an imposed environment and in the freely moving limb. The force field we measured in the interaction experiment was therefore comparable to the free movement in timing of muscle activity and also readily compared to the isometric force field, both in its timing and in the state of the musculoskeletal plant (Iwamoto *et al.* 1990). The different movement history experienced by the limb in the interaction experiment might influence the magnitude of neural muscle activation but it did not affect its timing. Furthermore, since the time of peak force and the active equilibrium was almost constant in different imposed environments, a set of samples at this time after an initial limb motion in a given environment could be regarded as a set of samples of an evolving force field in that environment.

Our results showed elements of trajectory generation that resemble those proposed in work on voluntary movements in humans (Sabes *et al.* 1998). In this study these elements are expressed in a force field context and, since our data represent the operations of the spinal cord alone, they incontrovertibly represent operation of elements of the segmental motor system.

In conclusion, the data here show that reflex proprioceptive feedback strongly regulates the muscle activation underlying behavioural force fields. These compensations may be described as precisely timed and systematic force field effects, provided a new dynamic force field primitive type is added. These additions are particularly important because it is possible that such motor primitives may have relevance for motor development, motor learning, and the cortical control of movement. It has been speculated that general purpose motor primitives or building blocks could be used for the voluntary regulation of posture and movement as well as for reflex behaviours (Bizzi *et al.* 1995). Both the postural primitives, the dynamic force field primitives and their timing, as described here, are well suited for this. However, it is important to note that the static data published previously and the measurements following movement that are introduced here represent isometric 'snapshots' of the force field regulation used in complex multi-joint limb motions and the spinal cord mechanisms that contribute to this. The task ahead is to understand the segmental and descending control of force field primitives throughout movement, in different environments, and to understand the circuitry that underpins these control processes.

Regarding motor learning, it has been shown that to approximate arbitrary force fields would require both conservative and non-conservative force field primitives. However, only conservative primitives were found in the frog. The dynamic primitives described here may allow a wider dynamic repertoire of fields while their transience and intrinsic timing are less challenging to limb stability. The timing may help minimize possible mechanical instability by always returning to a mechanically 'passive' elastic stability within a fixed time. Regarding equilibrium point models, it is notable that an equilibrium point exists in both the component force fields but only one has the spring-like properties associated with muscle. The convergence of the second dynamic force field is likely to be an additional aid to mechanical stability. The timing data here show the motor primitives can be linked to the spinal timing circuitry such as pattern generators. This timing may represent quantized reflex actions and choices. Finally, both the original postural field and the second field mechanism identified here are well suited to form primitives or components for the voluntary regulation of movement.

In summary, the data here show reflex proprioceptive feedback strongly regulates the muscle activation underlying behavioural force fields and these compensations may be described simply as precisely timed and systematic force field effects. However, it is important to note that the static data published previously and the measurements following movement that are described here represent isometric 'snapshots' of two aspects of the spinal cord mechanisms that contribute to the force field regulation used in complex multi-joint limb motions. The task ahead is to understand the control of force field primitives throughout movement, in different environments, and to understand the circuitry that underpins these control processes.

End of Case Study 4

Co-ordination

As we have seen, the effective co-ordination of movement and locomotion depends upon the information received by the CNS about the position, tension, etc. of the muscles. This information is provided by internal sense organs including muscle spindles, tendon organs, and joint receptors. These sense organs provide information about the relative positions of the limbs or other organs of movement. They also enable animals to make reflex responses to changes induced either by outside agents or by the movement of the animal itself.

Reflex behaviour is the simplest form of reaction to stimulation. Reflexes are normally automatic, involuntary, and stereotyped. A sudden change of tension in a muscle may result in automatic postural adjustment. A sudden change in the level of illumination may result in a reflex withdrawal response. Reflexes may be relatively localized, involving a single limb or other parts of the body. Sometimes, however, the whole animal may be involved in the reflex response. For example, the startle response of humans requires reflex co-ordination of numerous muscles. Similarly, the reflex withdrawal responses of invertebrates such as polychaete worms and various molluscs involve the whole body. In invertebrates, such reflexes are triggered by giant nerve fibres that carry very fast messages to all the muscles involved so that they contract suddenly and simultaneously.

Although normally automatic, most vertebrate reflexes can be subject to interference at the synapses that occur within the CNS. In the case of postural reflexes, there is often interference from other incompatible reflex mechanisms. When one reflex utilizes the same muscles as another, the reflexes are incompatible in the sense that both cannot occur simultaneously. Such pairs of reflexes are also neurologically incompatible in that stimulation of one reflex inhibits performance of the other. The inhibition is usually reciprocal so that one activity completely suppresses the other or is completely suppressed by the other. This type of reciprocal inhibition is typical of walking and other types of locomotion. It provides the most elementary form of co-ordination.

In general, muscular co-ordination is achieved by two main processes. These are central control and peripheral control. In the case of central control, precise instructions are issued by the brain and are obeyed by all the muscles involved. The co-ordination of swallowing movements in mammals seems to be of this type. Central control is important in the co-ordination of many skilled movements, which require rapid muscular activity. For example, the cuttlefish *Sepia* catches small crustaceans by means of two extensible tentacles, as illustrated in Figure 4.5. The control of attack falls into two phases. The first is a visually guided system in which the prey is brought into focus binocularly and movements of the prey are followed by movements of the cuttlefish in such a

120 Artificial ethology

Figure 4.5. *A cuttlefish catching a shrimp.*

way that visual error is reduced to zero. When this phase is complete, the tentacles are ejected suddenly and the prey is seized. This final phase is so rapid (about 30 ms) that there is no time for the tentacles to be guided visually onto the prey. The control mechanism is in two phases. The first phase involves a closed loop process by which visual feedback of the target position is compared with the aiming position. Only when these are equal is the visual error reduced to zero. The second phase is an open loop in that it involves no visual feedback. This can be demonstrated by turning off the lights during tentacle ejection, which does not affect prey capture. However, the prey is missed if it moves during ejection, showing that the cuttlefish is unable to correct its strike using visual feedback.

Locomotor rhythms are often generated centrally. Electrical recording from motor neurones and interneurones in insects has demonstrated central control rhythms for walking, swimming, and flight in arthropods. The precise patterns of behaviour, however, are often influenced by reflexes and feedback from the periphery. Thus, removal of proprioceptive input reduces stroke frequency in flying locusts, and wind stimulation increases motor neurone discharge. It appears that the fast walking movements of the cockroach are influenced less by peripheral stimuli than the slow walking movements of their relatives the stick insect (*Carausius*). Peripheral control of co-ordination is achieved through sense organs in the muscles and other parts of the body that send information to the brain and that thereby influence the instructions issued from the brain to the muscles. Peripheral control usually acts in co-operation with central control. For example, in the co-ordination of swimming movements in fish, the brain provides rhythmic signals that pass down the spinal cord in waves, co-ordinating the rhythmic movements of the fins and tail. In the dogfish (*Scyliorhinus*) the rhythm disappears if all nerves leading from the muscles to the brain are cut. If some nerves are left intact, however, the rhythm persists. Thus, it appears that some peripheral feedback is necessary to trigger the centrally produced rhythm. In cartilaginous fish (Chondrichthyes), including the dogfish, and all other sharks and rays, the fins show little independent rhythmic movement, but in bony fish (Teleostei) the fins can beat at different frequencies under some circumstances. As we saw in Chapter 1, von Holst (1939, 1973) showed that the rhythms of different fins can influence each other, a feature he called 'relative co-ordination'. Sometimes the rhythm of one fin attracts and dominates that of another so that they fall into step. In other cases, the amplitudes of fin movements summate so that the movements become smaller when the fins are out of step with each other and larger when they are in step.

In fishes and amphibians, locomotion appears to be mostly under the control of endogenous spinal rhythms. There is no evidence of specialized motor control exerted by the forebrain. Removal of the forebrain in fishes produces no change in posture or locomotion. In frogs and toads removal causes a general decrease in spontaneous movement, but electrical stimulation of the forebrain has no specific motor effects. The midbrain plays some role in motor control in fishes and amphibians. Electrical stimulation of the tectum in toads causes turning of the head and food snapping and swallowing. In higher vertebrates the higher brain is capable of exerting greater control of movement, but the automatic aspects of locomotion are still primarily controlled by the brain stem and spinal cord.

In mammals, the corticospinal tract is the most important pathway involved in the voluntary control of movement. It begins in the motor cortex and proceeds through the midbrain and brainstem to the spinal cord. This system, sometimes called the pyramidal system, is present in all mammals except the very primitive monotremes (e.g. platypus and spiny anteater). In the marsupial possum (*Trichosaurus*) the pyramidal axons run only to the midthorax, where they innervate the forelimbs. The hind-limbs are innervated by an extrapyramidal system.

The extrapyramidal system includes all non-reflex motor pathways not included in the corticospinal or pyramidal system. It is thought to be more primitive than the pyramidal system. In animals with little or no cerebral cortex, the basal ganglia of the extrapyramidal system are the most important centres of motor control. They are particularly well developed in birds, which have virtually no cerebral cortex, but a larger striatum than mammals. Thus, it appears that as the cerebral cortex developed, it brought into being a second source of motor co-ordination that acts through the pyramidal system.

In monkeys, neurones in various parts of the corticospinal pathways change their firing patterns during voluntary movements of the eyes, hands, and legs. Artificial stimulation of the motor cortex causes responses of single muscles or of single motor units within a muscle. More gross stimulation elicits discrete movements of whole limbs. It is possible to map the surface of the motor cortex according to the parts of the body moved in response to electrical stimulation. A similar map can be obtained for the sensory cortex in that the various parts of the body are represented differentially according to their sensory importance for individuals of the species concerned. Voluntary skilled movements are initiated via the cortical pyramidal neurones, while reflex maintenance of motor responses and of posture is controlled by nearby extrapyramidal neurones. There are estimated to be one million pyramidal neurones in humans.

Another part of the vertebrate brain that is important in co-ordination is the cerebellum. The cerebellum is not directly involved in postural reflexes or motor control. It serves as a monitor and co-ordinator of neural events involved in orientation and balance and in various other refined aspects of motor control. The basic neuronal organization of the cerebellum is similar in all classes of vertebrates, and the cerebellum has undergone less evolutionary change than any other part of the brain. The cerebellum receives information from the senses of vision, hearing, touch, equilibrium, and the condition of muscles and joints. It also has connections with the motor areas of the cerebral cortex. In addition to connections through the thalamus to the motor cortex, there are two-way connections to the sensory areas of the cortex. These are important in the maintenance of posture. Upright posture cannot be maintained without visual, vestibular, and proprioceptive information. The cerebellum combines visual and vestibular information about equilibrium with the state of contraction of the muscle spindles concerned. It sends appropriate instructions to the muscles, particularly through the gamma efferent fibres. The cerebellum thus exerts a modulating and refined control over the muscular contractions involved in the maintenance of posture and in the complex co-ordinated movements required for locomotion.

The locomotory behaviour of a given animal depends partly upon the anatomy, and partly upon the control system. These two factors interact in complex ways, and rather than making generalizations, it is informative to make an intensive study of a particular species, as is done in the next case study.

Case study 5: Robotic experiments on insect walking   *by Holk Cruse*

Hardware and software simulation of insect walking

It is known from the work of von Holst (1943) and Wendler (1964) that, in insects, the movements of individual legs are governed by independent control systems (for crustaceans see Chasserat and Clarac 1980). Each individual leg can step with its own rhythm. If the legs are only weakly coupled this leads to a behaviour von Holst called relative co-ordination. Subsequent investigations (Wendler 1966; for review, see Bässler 1993) showed that the whole system, that is, the leg and the accompanying neural control structures, forms a relaxation oscillator. During stance, the leg is on the ground, supports the body and, in the forward walking animal, moves backwards with respect to the body. The posterior transition point is called the posterior extreme position (PEP); it is determined in part by the position of the leg. At this transition, the behaviour switches from stance to swing. During swing, the leg is lifted off the ground and moved in the direction of walking to where it can begin a new stance; the anterior transition point is called the anterior extreme position (AEP).

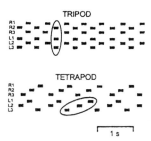

Figure CS 5.1. *The step patterns of a tripod and a tetrapod gait as produced by a stick insect.* The six traces represent the six legs. Black bars correspond to swing movement. Legs are designated as left (L) or right (R) and numbered from front to rear. Left and right legs on each segment (e.g. L1 and R1) always have a phase value of approximately 0.5. The phase value of adjacent ipsilateral legs (e.g. L1 and L2) is 0.5 in the tripod gait but differs in the tetrapod gait. (After Graham 1985.)

Although the legs are independent in principle, in the absence of strong disturbance during walking an insect typically shows a well-defined stepping pattern. The tripod gait — front and rear leg of one side swing together with the contralateral middle leg — is usually said to be typical for insects. Graham (1985) has shown in detailed investigations of stick insects that slowly walking animals, or animals walking under load, generally adopt the tetrapod gait (Figure CS 5.1). This appears to be also true for insects other than stick insects. The tetrapod gait can be described as a wave of swing movements travelling along the body from rear to front.

How does this co-ordination occur and how is it stabilized after disturbances? Results from stick insects and other arthropods show that the coupling between neighbouring legs can be described by simple rules (Cruse 1990).

In all, six different coupling mechanisms have been found in behavioural experiments with the stick insect. These are summarized in Figure CS 5.2a. One mechanism serves to correct errors in leg placement; another has to do with distributing propulsive force among the legs. These will not be considered here. The other four are used in the present model. The beginning of a swing movement, and therefore the endpoint of a stance movement, is modulated by three mechanisms arising from ipsilateral legs: (1) a rostrally directed inhibition during the swing movement of the next caudal leg (Figure CS 5.2b); (2) a rostrally directed excitation when the next caudal leg begins active retraction (Figure CS 5.2c); and (3) a caudally directed influence depending upon the position of the next rostral leg (Figure CS 5.2d). Influences (2) and (3) are also active between contralateral legs. The end of the swing movement in the animal is modulated by a single, caudally directed influence (4) depending on the position of the next rostral leg. This mechanism is responsible for the targeting

124 *Artificial ethology*

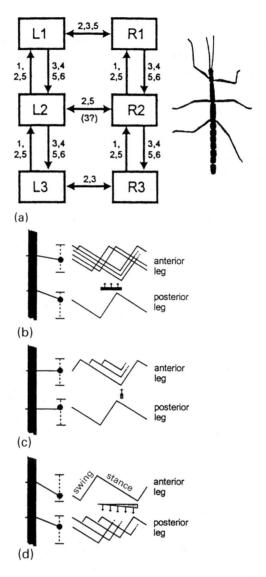

Figure CS 5.2. *Summary of the co-ordination mechanisms operating between the legs of a stick insect.* (a) The leg controllers are labelled R and L for right and left legs and numbered from 1 to 3 for front, middle, and hind legs. The different mechanisms (1 to 6) are explained in the text. (b), (c), (d) illustrate mechanisms 1, 2, and 3 respectively.

behaviour — the placement of the tarsus at the end of a swing close to the tarsus of the adjacent rostral leg.

These interleg influences are mediated in two parallel ways. The first pathway comprises the direct neural connections between the step pattern generators. The second pathway arises from the mechanical coupling among the legs. That is, the activity of one step pattern generator influences the movements and loading of all legs and therefore influences the activity of their step pattern generators via sensory pathways. This combination of mechanisms adds redundancy and robustness to the control system of the stick insect (Dean and Cruse 1995).

Simulation

Whether these interactions produce a sensible behaviour, can, however, not be made clear simply by intuition. Because of the complexity of the system a computer simulation is necessary to discover and understand the properties of this system. The simulations shown below demonstrate that these mechanisms deduced from behavioural experiments are sufficient to produce walking of the tripod or the tetrapod type and to stabilize the co-ordination against disturbances. This result means that the gaits are not explicitly calculated, but emerge from the co-operation of these local rules.

In the first simulations (Dean 1991a, b) a leg with only two degrees of freedom was considered. This can be moved forward/backward and up/down. The detailed geometry of the leg was subsequently taken into account in an extended version (Müller-Wilm et al. 1992). The geometry of the leg is shown in Figure CS 5.3. The coxa-trochanter and femur-tibia joints, the two distal joints, are simple hinge joints with one degree of freedom corresponding to elevation and extension of the tarsus, respectively. The subcoxal joint is more complex, but most of its movement is in a rostrocaudal direction around the nearly vertical axis. The additional degree of freedom allowing changes in the alignment of this axis is little used in normal walking, so the leg can be considered as a manipulator with three degrees of freedom for movement in

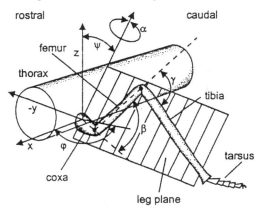

Figure CS 5.3. *Schematic model of a stick insect leg showing the arrangement of the joints and their axes of rotation.*

three dimensions. Thus, the control network must have at least three output channels, one for each leg joint.

Integrating new experimental results, the software simulation has recently been refined. This new simulation system, called Walk-net, is constructed using only simple, static artificial neurones. It does not apply Cartesian co-ordinates, but is based merely on intrinsic, egocentric co-ordinates like joint angles. Similarly, the control system consists of a number of distinct modules, which are responsible for solving particular subtasks. Some of them might be regarded as being responsible for the control of special microbehaviours. For example, a walking leg can be regarded as being in one of two states, namely performing a swing movement or a stance movement. These two microbehaviours are mutually exclusive. A leg cannot be in swing and in stance at the same time. Therefore, the control structure has to include a mechanism for deciding whether the swing or the stance module is in charge of the motor output. This is introduced here as a separate module called Selector-net. It is, however, an open question whether such a modular structure may also be found in the biological system.

To match experimental results that previously failed to demonstrate a robust central pattern generator producing strong intrinsic rhythms (Bässler and Wegner 1983), the Selector works on the basis of sensory input. The Selector-net consists of a two-layer feedforward net with positive feedback connections in the second layer. These positive feedback connections serve to stabilize the ongoing activity, namely stance or swing. This adds a kind of memory to the system, making it somewhat independent of small changes in the sensory input.

Control of the swing movement

The task of finding a network that produces a swing movement is simpler than finding a network to control the stance movement because a leg in swing is mechanically uncoupled from the environment and therefore, owing to its small mass, essentially uncoupled from the movement of the other legs.

A simple, two-layer feedforward net with three output units and six input units can produce movements that closely resemble the swing movements observed in walking stick insects (Cruse and Bartling 1995). The inputs correspond to three co-ordinates defining the actual leg configuration and three defining the target — the configuration desired at the end of the swing. In the simulation, the three outputs, interpreted as the angular velocities of the joints, are used to control the joints. The actual angles are measured and fed back into the net.

Through optimization, the network can be simplified to only eight (front and middle leg) or nine (hind leg) non-zero weights (Figure CS 5.4a). We believe this represents the simplest possible network for the task; it can be used as a standard of comparison with physiological results from stick insects. Despite its

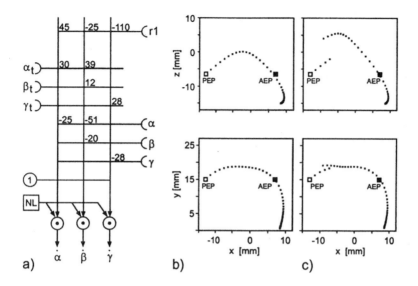

Figure CS 5.4. *A network for controlling swing movements.*

(a) The swing net. The motor units provide the angular velocities dα/dt, dβ/dt, and dγ/dt. The inputs provide the actual values of α, β, and γ as well as the angles of the target position ($α_t$, $β_t$, and $γ_t$). The numbers show the weights (x 100) used for the swing net of the left middle leg. (b, c) Swing movements produced by the net shown in (a). Side view (upper part) and top view (lower part) of the tarsus trajectory during a swing movement. The co-ordinate axes are explained in Figure CS 5.3. (b) A typical swing movement. (c) Simulation of an obstacle avoidance reaction: when the load receptor r1 is stimulated, the leg retracts and then continues protraction with a somewhat higher elevation.

simplicity, the net not only reproduces the trained trajectories (Figure CS 5.4b) but is able to generalize over a considerable range of untrained situations, demonstrating a further advantage of the network approach. Moreover, this Swing-net is remarkably tolerant with respect to external disturbances. The learned trajectories create a kind of attractor to which the disturbed trajectory returns. This compensation for disturbances occurs because the system does not compute explicit trajectories, but simply exploits the physical properties of the world.

The properties of this Swing-net can be described by the three-dimensional vector field in which the vectors show the movement produced by the swing net at each tarsus position in the workspace of the leg. Figure CS 5.5 shows the planar projections of one parasagittal section (left panel) and one horizontal section (right panel) through the workspace. The complete velocity fields are similar to the force fields discussed by Giszter (this volume) for the frog.

The weights of the Swing-net are chosen in such a way that the leg continues its movement with constant speed to some given distance below average ground

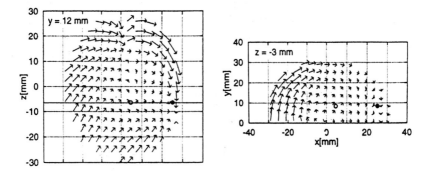

Figure CS 5.5. *Vector field representing the movement of the tarsus of a left front leg produced by the swing-net.* Left panel: Projection of a parasagittal section (y = 12 mm). Right panel: Projection of a horizontal section slightly below the leg insertion (z = −3 mm). Left is posterior, right is anterior. The average posterior extreme position (start of swing movement) and the average anterior extreme position (end of swing movement) are shown by an open circle and by a closed circle, respectively.

level. The velocity does not decrease as ground level is approached, even though ground contact may be anticipated. This agrees with the results of Cruse and Bartling (1995) where forces at ground contact were measured in a range between +5 mm and −5 mm relative to average ground level, but no dependency of force value on position was found.

The ability to compensate for external disturbances permits a simple extension of the Swing-net in order to simulate an avoidance behaviour observed in insects. When a leg strikes an obstacle during its swing, it initially attempts to avoid it by retracting and elevating briefly and then renewing its forward swing from this new position (Dean and Wendler 1982; Bässler et al. 1991). In the augmented Swing-net, an additional input similar to a tactile or force sensor signals such mechanical disturbances at the front part of the tibia (Figure CS 5.4a, r1) or the femur (Figure CS 5.4, r2). These units are connected by fixed weights to the three motor units in such a way as to produce the brief retraction and elevation seen in the avoidance reflex (Figure CS 5.4c). Other reflexes can be observed when the tibia is mechanically stimulated laterally or when the femur is touched dorsally. These reflexes have been implemented in an analogous manner (Figure CS 5.4).

In the model, the targeting influence reaches the leg controller as part of the input to the Swing-net (Figure CS 5.4). These signals can be generated by a simple feedforward net with three hidden units and logistic activation functions (Figure CS 5.6, Target-net) which directly associates desired final joint angles for the swing with the current joint angles of a rostral leg, in such a way that the tarsus of the posterior leg is moved in the direction of that of the anterior leg. Compared to a first version, the new target net has direct connections between the input and the output layer. There is no explicit calculation of the position of either tarsus. Physiological recordings from local and intersegmental interneurones (Brunn and Dean 1994) support the hypothesis that a similar approximate algorithm is implemented in the nervous system of the stick insect.

In Walk-net, it is assumed that the end of the swing is determined by a

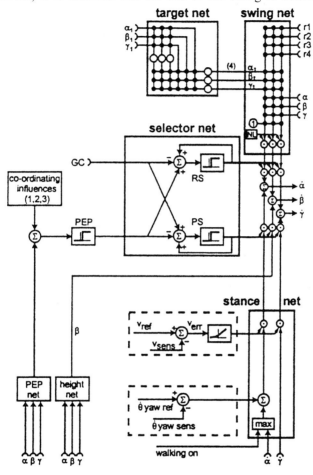

Figure CS 5.6. *The leg controller.* The controller consists of three parts: the swing net, the stance net, and the selector net which determines whether the swing net or the stance net controls the motor output (the velocity of the three joints α, β, and γ). The selector net contains four units: the PEP unit signalling posterior extreme position, the GC unit signalling ground contact, the RS unit controlling the return stroke (swing movement), and the PS unit controlling the power stroke (stance movement). The target net transforms information about the configuration of the anterior, target leg (α_1, β_1, and γ_1) into angular values for the next caudal leg which will place the two tarsi close together. These desired final values (α_t, β_t, and γ_t) and the current values (α, β, and γ) of the leg angles are input to the swing net together with a bias input (1) and four sensory inputs (r1 - r4) which are activated by obstructions blocking the swing and thereby initiate different avoidance movements. A non-linear influence (NL) modulates the velocity profile. For details see Cruse *et al.* (1996).

sensory stimulus affecting the tarsal mechanoreceptors. (Load receptors most probably play an important role, too, but are not yet taken into account in Walk-net.) As soon as this stimulus is above a given threshold, the selector net is forced to switch from swing to stance. In the animals, however, the situation is not as simple. This may not be unexpected because any arbitrary mechanical contact in the early swing, for example, should be considered an obstacle rather than a stimulus to finish swing. So one might assume that there is some internal state, an 'expectation value' or conversely, a value for 'swing motivation', which tells the system whether a given stimulus should be treated as an obstacle or as ground contact.

How do the animals behave? When a possible substrate, for example a round wooden rod, is held at different positions in the swing trajectory of a leg (Figure CS 5.7a), the probability that it will be grasped by the tarsus to end the swing is very low when the rod is in the caudal 40% of the step range, but the probability increases about linearly to 100% as the position approaches that of the normal anterior extreme position, or AEP (Figure CS 5.7b). This raises the question of how this decreasing 'swing motivation' is produced? Experimental results showed that it is not simply the position of the leg because the results for a given leg position are different when the swing duration is artificially prolonged by about 500, 750, or 1000 ms. This result also shows that another possible parameter, namely time elapsed since the beginning of the swing, cannot be the only relevant parameter, because these prolonged swing movements far exceeded the duration of a normal swing (150 to 200 ms). If time were the only parameter, the probability of ending the swing should have been high for all three cases.

A third experiment hinted at a solution. One can easily manipulate the position of a leg of an insect walking on a treadmill by placing this leg on a platform fixed beside the treadwheel. If this platform is held far enough in the anterior range of leg movement, this leg remains standing on the platform, while the other five legs continue to walk. There are reasons to believe that this leg is in a walking rather than in a standing mode (Schmitz 1985). When the basic experiment is repeated with a middle leg, and the position of the ipsilateral front leg is manipulated as described, the middle leg shows reactions that not only depend on its own position, but also on that of the front leg behind its normal AEP. These results suggest that it is the distance between front and middle leg that determines the 'motivation' for the middle leg to accept a stimulus as ground contact, a hypothesis that can quantitatively describe all our data.

How might the probabilistic property found in these experiments be introduced into the network model? The simple static neurones used in Walk-net up to now cannot do this. However, this behavioural reaction can be described most simply by the following assumption (Figure CS 5.7c): an interneurone provides a noisy measure of the distance between front and middle leg. The sum of both is rectified such that no negative excitations can occur at the output. This value inhibits the sensory input from the tarsal receptors. In this way the probability for a tactile input to exceed threshold and elicit a grasping reaction increases with the activity level of the neurone representing the distance.

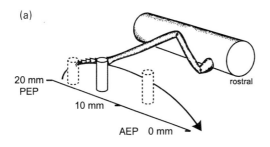

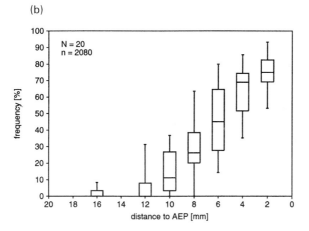

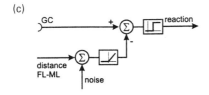

Figure CS 5.7. *Accepting a stimulus as ground contact to end swing.* (a) During swing movement, a wooden stick is held in the tarsus trajectory at different positions. (b) The dependency of the frequency of the grasping reaction of the middle leg tarsus on the test substrate's position relative to its normal AEP (distance in mm). (c) A simple circuit which can describe the results and may replace the simple ground contact (GC) input to the selector net shown in Figure CS 5.6. For further explanation see text.

At least qualitatively this result fits those of Brunn and Dean (1994) and Brunn (1998), who found an interneurone that increases its activity the closer both neighbouring legs approach each other. This neurone in turn inhibits the protractor motor neurones and excites the retractor motor neurones, and thus can contribute to the transition from swing to stance.

Control of the stance movement and co-ordination of supporting legs

The control of the stance movement appears to be more difficult. It is not enough simply to specify a movement for each leg on its own as in the case of the swing movement: the mechanical coupling through the substrate means that efficient locomotion requires co-ordinated movement of all the joints of all the legs in contact with the substrate, that is, a total of 18 joints when all legs of an insect are on the ground. The task is a non-linear problem, because we have to deal with rotational movements. This is even more the case when the rotational axes of the joints are not orthogonal, as is often the case for insect legs and for the basal leg joint in particular. Another non-linearity arises from the fact that the number and combination of mechanically coupled joints varies from one moment to the next, depending on which legs are lifted.

For straight walks, one could simplify the problem by assuming that the trajectories of the leg endpoint follow a straight line parallel to the long axis of the body. This assumption, however, is only approximately the case in normal walking. It definitely does not hold when the animal negotiates a curve, which requires the different legs to move along different arcs at different speeds. Superimposed on all these problems we have the above mentioned problem that the control system has to deal with extra degrees of freedom. This means that the control system has to decide, by applying some criteria, which of the possible solutions should be selected.

In machines, these problems can be solved by traditional, though computationally costly, methods, which consider the ground reaction forces of all legs in stance and seek to optimize some additional criteria, such as minimizing the tension or compression exerted by the legs on the substrate. Owing to the nature of the mechanical interactions and inherent in the search for a globally optimal control strategy, such algorithms require a single, central controller; they do not lend themselves to distributed processing. This makes real-time control difficult, even in the still simple case of walking on a rigid substrate. The much smaller bandwidth and the much slower computational speed of the biological systems compared to the technical ones makes real-time control even more difficult.

Further complexities arise in more irregular, natural walking situations, making solution difficult even with high computational power. These problems arise, for example, when an animal or a machine walks on a slippery surface or on a compliant substrate, such as the leaves and twigs encountered by stick insects. Any flexibility in the suspension of the joints further increases the degrees of freedom that must be considered and the complexity of the computation. Further problems for an exact, analytical solution occur when the length of leg segments changes during growth or their shape changes through injury. In such cases, knowledge of the geometrical situation is incomplete, making an explicit calculation difficult, if not impossible. Such problems already arise during normal walking: the positions and orientations of the axes in the non-rigid joints may change due to load changes elicited by different orientation with respect to gravity.

Local positive feedback – the solution?

Despite the evident complexity of these tasks, they are mastered even by insects with their 'simple' nervous systems. Hence, there has to be a solution that is fast enough that online computation is possible even for slow neuronal systems. How can this be done? Several authors (e.g. Brooks 1991) have pointed out that some relevant parameters do not need to be explicitly calculated by the nervous system because they are already available in the interaction with the environment. This means that, instead of an abstract calculation, the system can directly exploit the dynamics of the interaction and thereby avoid a slow, computationally exact algorithm. To solve the particular problem at hand, we propose to replace a central controller with distributed control in the form of local positive feedback (Cruse *et al.* 1996). Compared to earlier versions (Müller-Wilm *et al.* 1992), this change permits the stance net to be radically simplified. The positive feedback occurs at the level of single joints: the position signal of each is fed back to control the motor output of the same joint (Figure CS 5.6, Stance-net). How does this system work? Let us assume that any one joint is moved actively. Then, because of the mechanical connections, all other joints begin to move passively, but in exactly the proper way to maintain the mechanical integrity of the system. Thus, the movement direction and speed of each joint does not have to be computed because this information is already provided by the physics. The positive feedback then transforms this passive movement into an active movement.

This idea is supported by an earlier finding of Bässler (1976) showing a reflex reversal for the femur-tibia joint which could be interpreted as a positive feedback. Therefore, we decided to implement this solution for the control of the stance movement. There are, however, several problems to be solved. The first is that positive feedback using the raw position signal would lead to unpredictable changes in movement speed, not the nearly constant walking speed that is usually desired. This problem can be solved by introducing a kind of bandpass filter into the feedback loop. The effect is to make the feedback proportional to the angular velocity of joint movement, not the angular position.

In the simulation, this is done by feeding back a signal proportional to the angular change over the preceding time interval.

The second problem is that using positive feedback for all three leg joints leads to unpredictable changes in body height, even in a computer simulation neglecting gravity. In the stick insect, body height is controlled by a distributed system in which each leg acts like an independent, proportional controller (Cruse 1976). However, maintaining a given height via negative feedback appears at odds with the proposed local positive feedback for forward movement. How can both functions be fulfilled at the same time? To solve this problem we assume that during walking positive feedback is provided for the α joints and the γ joints (Figure CS 5.6, Stance-net), but not for the β joints. The β joint is the major determinant of the separation between the substrate and the leg insertion in the body, which determines body height.

A third problem inherent in using positive feedback is the following. Let us assume that a stationary insect is pulled backwards by gravity or by a brief tug from an experimenter. With positive feedback control as described, the insect should then continue to walk backwards even after the initial pull ends. This has never been observed. Therefore, we assume that a supervisory system exists which is not only responsible for switching on and off the entire walking system, but also specifies walking direction (normally forward for the insect). This influence is represented by applying a small, positive input value (Figure CS 5.6, 'walking on') which replaces the sensory signal if it is larger than the latter (the box 'max' in Figure CS 5.6, Stance-net).

To permit the system to control straight walking and to negotiate curves, a supervisory system was introduced which, in a simple way, simulates optomotor mechanisms for course stabilization that are well known from insects and have also been applied in robotics. This supervisory system uses information on the rate of yaw, such as visual movement detectors might provide. It is based on negative feedback of the difference between the desired turning rate and the actual change in heading over the last time step. The error signal controls additional impulses to the α joints of the front and hind legs, which have magnitudes proportional to the deviation and opposite signs for the right and left sides. In earlier versions, this bias was given to the front legs only. A much better behaviour can be found when the bias is also given to the hind legs. With this addition and the yaw reference set to zero, the system moves straight (Figure CS 5.8a) with small, side-to-side oscillations in heading such as can be observed in walking insects. To simulate curve walking (Figure CS 5.8b), the reference value is given a small positive or negative bias to determine curvature direction and magnitude.

Finally, we have to address the question of how walking speed is determined in such a positive feedback controller. Again, we assume a central value which represents the desired walking speed. This is compared with the actual speed, which could be measured by visual inputs or by monitoring leg movement. This error signal is subject to a non-linear transformation and then multiplied with the signals providing the positive feedback for all α and γ joints of all six legs (Figure CS 5.6, Stance-net).

As a first step, i.e. before the planned implementation in a real robot, the behaviour of Walk-net was tested quantitatively in a software simulation. As the control principle relies on the existence of a body, the mechanics of the body had to be simulated, too. For this purpose, a recurrent network, called MMC, was developed, which will not be explained here. It suffices to say that it consists of a 78 unit recurrent network structured like a Hopfield net (Hopfield

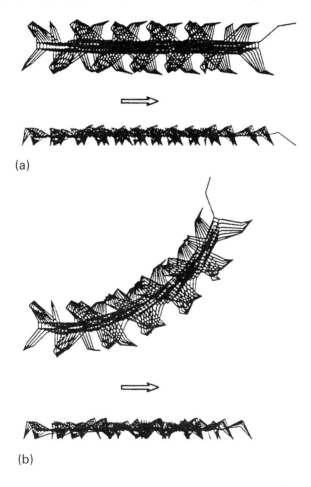

Figure CS 5.8. *Simulated walk by the basic six-legged system.* Negative feedback is applied to all six β joints and positive feedback to all α and γ joints as shown in Figure CS 5.6. Movement direction is from left to right (arrow). Leg positions are illustrated only during stance and only for every second time interval in the simulation. Each leg makes about five steps. Upper part: top view. Lower part: side view. (a) Straight walking, (b) curved walking.

1982) augmented by some non-linear elements. The network represents the configuration of the central body and the six legs which each consist of three elements (coxa, femur, tibia, see Figure CS 5.3). The network can adopt a state representing any geometrically possible position of this 19 body system. After a not too large external disturbance the system always relaxes to a state that corresponds to a geometrically possible configuration. In this way, it behaves like a mechanical device in which the joints are provided with adjustable springs. The new position depends on the old body position and the influence of the disturbance. The output values of the Walk-net controller are then considered as such disturbance influences and in this way move the simulated body. This is a purely kinematic simulation; i.e. no inertia and no forces are taken into account.

Local positive feedback solves these as well as further problems

As is shown in Figure CS 5.8a for the case of straight walking, this network is able to control proper co-ordination. Steps of ipsilateral legs are organized in triplets forming 'metachronal waves', which proceed from back to front, whereas steps of the contralateral legs on each segment step approximately in alternation. With increasing walking speed, the typical change in co-ordination from the tetrapod to a tripod-like gait is found. For slow and medium velocities the walking pattern corresponds to the tetrapod gait with four or more legs on the ground at any time and diagonal pairs of legs stepping approximately together; for higher velocities the gait approaches the tripod pattern with front and rear legs on each side stepping together with the contralateral middle leg. The co-ordination pattern is very stable. For example, when the movement of one leg is interrupted briefly during the power stroke, the normal co-ordination is regained immediately at the end of the perturbation. Furthermore, the model can cope with obstacles higher than the normal distance between the body and the substrate. It continues walking when a leg has been injured (Cruse *et al.* 1998).

What about curve walking? The typical engineer's solution is to determine the curve radius and the centre of the curve. With these values the trajectories of the different legs are calculated and then, using inverse kinematics, the trajectories for the joint angles are determined. In our case, too, a value is required to determine the tightness of the curve. This, however, does not need quantitatively to correspond to the curve radius. The value is only used as an amplification factor for the feedback loop of front and hind legs. This value can deliberately be changed from one moment to the next. No further calculations are necessary. The introduction of the local bandpass filtered positive feedback in 12 of the 18 leg joints provides a control system which as far as we can see cannot be further simplified, because it is decentralized down to the level of the single joints. This simplification has the side effect that computation time can be minimized. The

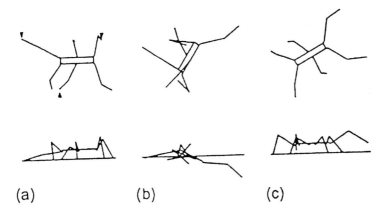

Figure CS 5.9. *Righting behaviour.* (a) By clamping the tarsi to the ground (arrowheads), the system is made to fall leading to (b) a disordered arrangement of the legs. Nevertheless, the system stands up without help (c) and resumes proper walking.

essential advantage, however, is that, by means of this simplification and the consideration of physical properties of the body and the environment, all problems mentioned above can easily be solved, although they, at first sight, seemed to be very difficult.

Unexpectedly, the following interesting behaviour was observed. A massive perturbation, for example by clamping the tarsi of three legs to the ground, can make the system fail (Figure CS 5.9). Although this can lead to extremely disordered arrangements of the six legs, the system was always able to stand up and resume proper walking without any help. This means that the simple solution proposed here also eliminates the need for a special supervisory system to rearrange leg positions after such an emergency.

One major disadvantage of our simulation is its pure kinematic nature. To test the principle of local positive feedback at least for straight walking, we have performed a dynamic simulation for the six-legged system under positive feedback control during stance. The basic software was kindly provided by F. Pfeiffer, TU Munich. No problems occurred. Nevertheless, a hardware test of the walking situations is necessary, because all these software simulations may lack an important, but yet overlooked influence. Such a test is currently being performed in collaboration with M. Frik, at the University of Duisburg, and his robot TARRY (Figure CS 5.10) (Frik *et al.* 1998).

Conclusion

It might be thought that, considering all the problems a walking system has to deal with under seminatural conditions, a high motor intelligence would be necessary for the control of walking. We have, however, seen that the control system neither has to be very complicated nor does it require a centralized architecture. On the contrary, the seemingly most difficult problems are solved by the structurally most simple subsystems. This simplification is possible because the physical properties of the system and its interaction with the world are exploited, instead of using an abstract, explicit computation. No explicit internal world model is required. Thus, 'the world' is used as 'its own best model' (Brooks 1991d). This principle is implemented at several places in the control system.

(1) The quasi-rhythmic leg movements are not produced by an endogenous central oscillator. Instead they result from the interaction of the neuronal control system and the environment.

(2) No explicit computation of the complete trajectory is necessary for the generation of the swing movement. Instead the instantaneous continuation of a movement is determined on the basis of the current sensory input values.

(3) The discrepancy between the complexity of the task and the simplicity of the solution is most obvious in the case of the control of the stance movement.

Furthermore, the Walk-net simulation shows that simple local rules can produce unexpected properties at the level of the whole system: (1) the four local co-ordinating mechanisms produce tripod or tetrapod gaits that are stable in the face of disturbances, and (2) a combination of positive and negative

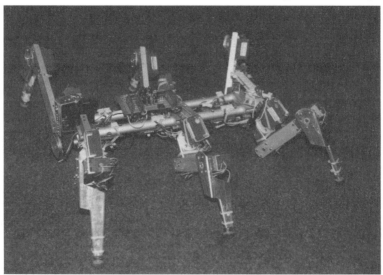

Figure CS 5.10. *The robot TARRY.*

feedback permits the control system to maintain body height and to right the body after a fall.

None of these results could have been obtained without the use of computer simulations. However, there remain important questions that could be answered only by hardware simulation, i.e. using a real robot, such as that illustrated in Figure CS 5.10. Many behavioural properties are a function, not of the brain alone, but of the brain, body, and environment in combination. Thus we must somehow include the body and the environment in the simulation. (In our case this is relevant to the question of whether high-pass filtered positive feedback might really solve the problems addressed above.) One of the best ways of doing this is to build a real robot.

Although it may seem that a software simulation is sufficient, it may sometimes be cheaper to build a real robot, and this approach may also be less prone to errors. Realistic software simulation of a real-life situation may be very complex. For example, simulation of the kinematics of the 19-segmented body with its high number of degrees of freedom is a time-consuming endeavour, even when restricted to walking in a straight line. Moreover, in such a complex simulation, there are inevitable over-simplifications, and some aspects of the real situation may be overlooked. Such inaccuracies cannot occur in a hardware simulation. Dealing with reality directly tends to make hidden questions obvious, and may open up many new questions.

End of Case Study 5

Underwater locomotion

Underwater locomotion is of particular interest, because it generally takes place in a current, thus posing additional problems. The bodyshape of lobsters, for example, is designed to exploit hydrodynamic principles. When the lobster faces into a current it gains stability from the flow of water over its specially shaped shell. This enables the lobster to achieve impressive mobility by walking forwards, backwards, or sideways, as we see in the next case study, which neatly demonstrates the applicability of biological analysis to engineering design.

Case study 6: Building a robotic lobster
by Joseph Ayers

The aim of this project is to produce robots that are truly biomimetic. They will use actuators that mimic muscle, sensors based on hair-like microcantilevers, and controllers based on models of the organization of central networks (Ayers *et al.* 1998). To capture the hydrodynamic advantages of the model systems we

140 Artificial ethology

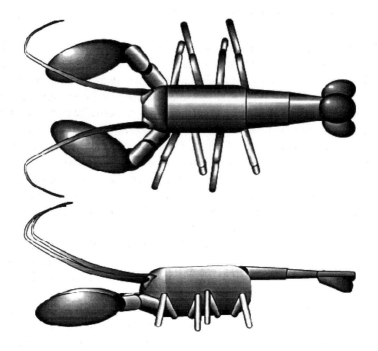

Figure CS 6.1. *Proposed lobster-based robot.*

are adapting, as closely as possible, the actual morphology of the model animals in the construction of the robots. An artist's impression of the proposed robot is illustrated in Figure CS 6.1.

In developing these systems we use a synthetic approach which combines kinematic analysis of animal behaviour and neuronal network modelling with robotic design and fabrication. We use the animals as a literal blueprint of the desired design and iterate towards a robotic design which implements as much as possible of the animal control systems, mechanics, and hydrodynamics. The design process requires a complete solution. These systems are intended for remote sensing in marine environments and therefore require, at least, supervised autonomy. Although much is known about the orientational and taxic behaviour of crustaceans (Sandeman and Atwood 1982), there has been no attempt, to date, to formulate a complete ethogram of the species. The design of an autonomous robot requires such an ethogram and we are closely linking the robotic design process with the quantification of behavioural sequencing exhibited by lobsters adapting to the marine environment.

In what follows I shall review the organization of crustacean motor systems, illustrate how this organizational scheme can be implemented in a biomimetic robot and describe how this control architecture has led to the development of tools to establish a state-based ethogram of lobster behaviour.

Organization of crustacean motor systems

The central neuronal mechanisms underlying locomotion were first established by the study of simple animals including decapod crustacea, insects, and annelids (Hoyle 1976; Kennedy and Davis 1977). These mechanisms have been formalized into a general model, the command neurone, co-ordinating neurone, central pattern generator model (CCCPG model) which has been demonstrated to be conserved throughout the invertebrates (Herman *et al.* 1976; Kennedy and Davis 1977; Stein 1978; Evoy and Ayers 1981). The model is composed of five major classes of components including central pattern generators (Pinsker and Ayers 1983; Selverston and Moulins 1987), command systems (Kupferman and Weiss 1978), co-ordinating systems (Stein 1976), proprioceptive and exteroceptive sensors (Wiersma and Roach 1977), and phase- and amplitude-modulating sensory feedback (Stein 1978).

The fundamental governing concept of this model is that the motor output that underlies behaviour is generated by genotypically specified central pattern generators which are modulated by peripheral exteroceptive and proprioceptive feedback during behaviour (Delcomyn 1980). In other words, the central nervous system can generate central motor programs in the absence of sensory feedback. This central pattern generation model differs fundamentally from reflex-chain models where sensory feedback is necessary to specify transitions between different phases of a cyclic behaviour (Sherrington 1906). During rhythmic behaviour, feedback from proprioceptors will be rhythmic, thus the role of sensory feedback might be best considered in the context of coupled oscillators (Pinsker and Ayers 1983).

The central component consists of segmental central pattern generators (CPGs) which control the motor neurones and muscles of each limb, co-ordinating systems which determine the phase relations or gaits between the CPGs of different limbs, and command systems which specify and modulate the behaviour generated by the CPGs. The command systems represent the control locus at which the decision to generate a particular behaviour is made (Kupferman and Weiss 1978; Eaton and DiDomenico 1985). The peripheral component consists of exteroceptive and proprioceptive sensors (Wiersma and Roach 1977) which generate adaptational reflexes at the command level (exteroceptive or orientational reflexes), the CPG level (phase-modulating reflexes), and/or the motor neurone level (amplitude-modulating reflexes).

The neuronal control mechanisms of insects and decapod crustaceans (cockroaches, locusts, lobsters, crayfish, and crabs) have been subjected to considerable reverse engineering over the past 25 years, adequate to permit robust synthetic models of their underlying organization (Beer 1991). In several cases the actual synaptic networks have been established by electrophysiological stimulation and recording (Pearson 1972; Chirachri and Clarac 1989). These biological models can be readily adapted to robotic control (Ayers and Crisman

1992; Beer *et. al.* 1992; Brooks 1992). Biologically based reverse engineering is a very effective procedure both in designing autonomous underwater robots as well as in establishing detailed higher order control schemes for remote sensing procedures. The control schemes by which a lobster searches for and acquires prey provide excellent solutions to the problem of how an underwater robot can successfully search for and investigate underwater objects.

Organization of the lobster walking system

The control of the decapod walking system has been investigated at almost all levels, from the behavioural to the neuromodulatory. Combined with comparative data from other oscillatory systems in the crustacean CNS, it is possible to construct reasonably complete control structures based entirely on known components (Evoy and Ayers 1981; Ayers and Crisman 1992).

Limb movements Lobster walking movements are performed around three major limb joints (Figure CS 6.2): the thoraco-coxal joint generates protraction and

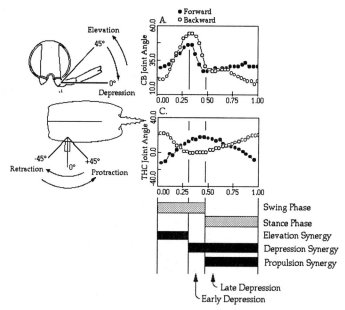

Figure CS 6.2. *Joint movement and muscle synergy control patterns.* The two panels at the left indicate the angular co-ordinates of movement of the coxo-basal joint (elevation and depression) and the thoraco-coxal joint (protraction and retraction). The two upper graphics on the right indicate the angles of the coxo-basal and thoraco-coxal joints plotted against phase in the step cycle. Forward walking is indicated by closed circles while backward walking is indicated by open circles. The lower right panel indicates the duration of the swing phase and stance phase (stippled bars) and the corresponding elevation, depression, and propulsion phases.

retraction movements, the coxo-basal joint generates elevation and depression movements while the mero-carpopodite joint generates extension and flexion movements (Ayers and Davis 1977a). Cyclic elevation and depression movements of the coxo-basal joints underlie the swing and stance phase movements respectively for walking in all four directions (Figure CS 6.2). Propulsive forces are generated by movements of the thoraco-coxal and mero-carpopodite joints. Propulsive forces that are synergistic with depression for walking in one direction become antagonistic for walking in the opposite direction (Figure CS 6.2). The lobster step cycle consists of three phases: an early swing phase lifts the limb tip towards the initial position of the stance, an early stance phase drops the leg to initiate the stance, and during the late stance phase the limb applies propulsive force and compensates for gravity. The swing phase is always constant in duration and variation in the step period is mediated by variation in the duration of the stance phase (Ayers and Davis 1977a).

Crustacean command neurones One of the primary advantages of using lobsters and crayfish for neuronal modelling and simulation is the relative parsimony of the central nervous system. Much of the recognizable behaviour of macruran crustacea can be evoked by selective electrical stimulation of single neurones descending through the circumoesophageal connectives from the brain to the ventral thoracic and abdominal ganglionic chain (Bowerman and Larimer 1974a, b). These commands can be divided into trigger commands where the response outlasts the stimulus and gate commands where tonic stimulation can evoke relatively stable states of rhythmic behaviour (Bowerman and Larimer 1974b; Stein 1978).

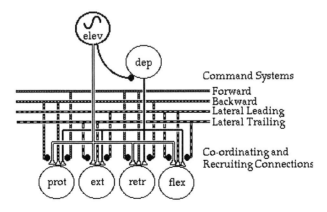

Figure CS 6.3. *Hypothetical neuronal network for the control of omnidirectional stepping.* Closed circles indicate presynaptic or conventional inhibitory connections. Open triangles indicate excitatory connections.

Neuronal circuits controlling locomotion in crustacea The bulk of the macruran behavioural repertoire can be evoked by stimulation of single descending command neurones (Bowerman and Larimer 1974a, b; Hoyle 1976). A hypothetical neuronal network for the control of walking based on neuronal oscillators, command and co-ordinating neurones has been developed (Figure CS 6.3). This simple network is capable of controlling walking in four directions (Ayers and Davis 1977; Ayers and Crisman 1992). The basis of the network is the use of direct connections between oscillator neurones and propulsive force neurones (Chirachri and Clarac 1989) modulated by command systems (Cattaert *et al.* 1990). According to this model, command systems specify the co-ordination pattern through the modulation of particular connections to specify walking direction (Figure CS 6.3). Command systems also specify the amplitude of movements by controlling recruitment from the motor unit pool (Davis and Kennedy 1972; Ayers and Crisman 1992).

Lobster-based ambulation controller

At present, we have completed the low-level components of the omnidirectional ambulation controller, and we are currently implementing the controls for behavioural sequences. We implemented the CCCPG model for omnidirectional walking as a finite state machine. The ambulation controller program generates digital output for actuator control, and also provides real-time chart displays of motor programs imitating electromyograms and joint movements. The program treats the CCCPG model components as objects that pass messages relating to status changes. Changes in walking direction, speed, load, etc. are effected by menu selections or key strokes. The chart displays generated by the program are

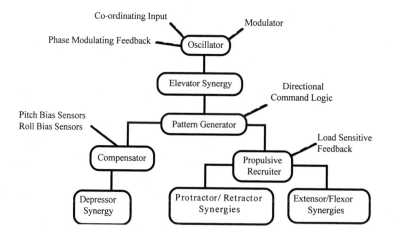

Figure CS 6.4. *Message hierarchy of the state machine.* State change messages pass down the hierarchy.

the basis of many of the figures in this report showing experimental results.

At the single limb control level, the ambulation controller relies on three major classes of components which control the elevator, depressor, protractor, retractor, extensor, and flexor synergies (Figure CS 6.4). The oscillator component is a software clock which regulates the period of stepping as well as the duration of the swing or elevator phase fraction of the stepping cycle. The clock maintains step timing registers which contain the parameter associated with the desired stepping period. At the termination of each step cycle the oscillator loads the step timing registers with the clock tick values associated with the expected end of the elevator phase, and with the end of the step cycle based on the desired stepping period. During ongoing operation, the oscillator continuously compares the processor clock ticks with the two registers, and, when the target times are achieved, issues state transition messages to the co-ordinator, the recruiter, and the neuronal oscillators of any ipsilateral or contralateral governed limbs.

The second major component of the ambulation controller is the co-ordinator which determines the pattern of discharge of bifunctional synergies. The co-ordinator responds to the state transition message and desired period parameter from the oscillator, polls the walking command logic, and determines through a truth table which synergies should be active, and sets or clears the Boolean values associated with different bifunctional synergies. The truth table implements the presynaptic inhibitory logic of the neuronal circuit model and specifies the excitatory connections that would be disabled by the directional command.

The third major component of the ambulation controller is the recruiter which determines which of the elements of the propulsive force synergies are active (Figure CS 6.5). The recruiter responds to the desired period parameter as well as load-sensitive feedback and sets or clears the Boolean values associated with different elements of the active propulsive pool. In the current implementation, each unit within a synergy is represented as a Boolean value (on or off) although it could easily be represented as a scalar variable to provide more detailed amplitude control.

Figure CS 6.5 shows a graphical realization of the output of the ambulation controller demonstrating adaptation to speed during forward and backward walking. The output is realized as a strip-chart recording of the timing of activity in the different synergies in the same context as electromyograms or nerve recordings. In propulsive force synergies the width of the trace indicates the degree of recruitment within the pool. Note that during increases in speed the period of stepping decreases while propulsive force synergies are selectively recruited.

In addition to controlling forward and backward walking, the ambulation controller implements omnidirectional locomotion including lateral walking on the leading and trailing sides (Ayers and Davis 1977). To complete this

146 Artificial ethology

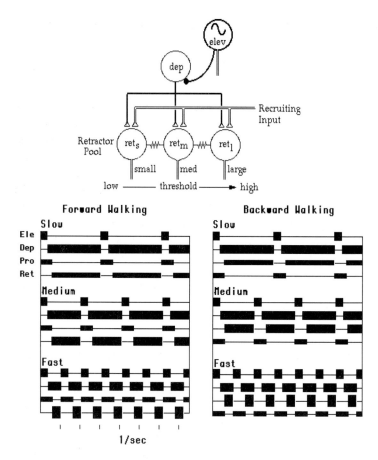

Figure CS 6.5. *Motor programs for unidirectional walking generated by the finite state machine.* The recruitment circuit (above) mediates size-ordered recruitment of actuators. The graphical representations show adaptation to speed indicating size-ordered recruitment of propulsive synergies.

orthogonal symmetry, we have included directional and recruiting logic for backward, lateral leading, and lateral trailing walking. In all cases, the directional logic eliminates the inappropriate synergies by decoupling propulsive force synergies from the elevation/depression oscillation and adds speed-dependent recruitment to the propulsive force synergy. The added connections include those that decouple elevation from flexion and depression from extension during lateral leading walking and those that decouple elevation from extension and depression from flexion during lateral trailing walking. The output of the finite state machine is a set of control signals that specify the timing and amplitude of actuator action. Movement synergies that are synergistic for walking in one direction become antagonistic for walking in the opposite direction. Adaptation to speed is mediated by increases in the frequency of the pattern as well as recruitment of the propulsive synergy (Ayers and Crisman 1992).

Sensors

Crustacean sensors code environmental parameters in terms of labelled lines (Bullock 1978). Each sensor is represented by an array of labelled line elements, each of which codes for a particular sensory modality (gravity, water current, etc.) as well as receptive fields (i.e. orientation relative to horizontal, water currents from the front, rear or the sides, etc.). In the controller implementation, sensor objects pass messages to command objects based on the actual input. Thus the message contains the modality and orientation and evokes different methods from the command system, based on these parameters. During locomotion in currents and surge, lobsters predominantly rely on three sensors, the antennae, water current receptors, and the statocysts or vestibular receptors. Although the actual biological sensors are complex, they can be readily modelled by mechanical transducers to code environmental information in the same fashion as the lobster nervous system.

Antennae
Real lobsters use their antennae extensively both as tactile receptors for collision avoidance as well as water current receptors. They are also controlled by visual input to participate in the antennal pointing reflex. In a robotic implementation they require both active motor control as well as a set of labelled lines which code for contact as well as current forces. In the simplest implementation, the antennae should be constructed of a tapered tube of moderate restoring force. Antagonist actuators would mediate protraction and retraction in the yaw plane. Strain gauges placed at the distal end would code for contact while others placed at the basilar portion would code for water currents as well.

Water current receptors
Crustaceans detect water currents by a combination of antennae as well as specialized directional sensors called hair fan organs which are distributed over the carapace and thorax. (Laverack 1962). The hair fan organs consist of a cluster of hairs which can pivot in one plane (Figure CS 6.6). Water currents deflect hair pegs where the pivot of the hairs is oriented at right angles to the direction of current, so the organs provide labelled lines which can code both amplitude and direction. In a robotic implementation, a hair fan analogue could be built using a tiny strain gauge (Figure CS 6.6) mounted inside an open tube; a water current sensing system could be implemented by mounting several such sensors in different orientations on the robot's 'claws'.

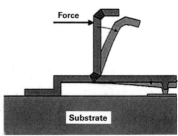

Figure CS 6.6. *Biomimetic water current sensors.* Left panel: Hair fan organ of macruran decapods. (After Laverack 1962.) The central hair cluster is deflected by currents to provide labelled line outputs for currents in different directions. Right panel: Biomimetic sensor structure. The extent to which the structure is deformed by a current could be measured by a strain gauge. Each such element would be mounted inside a small tubular housing, which would constrain the element to respond to currents in a particular orientation, and would protect the gauge from mechanical insult. The individual sensors would be placed on the 'cheliped' hydrodynamic control surfaces at different angles. Simple thresholding of the strain gauge responses would indicate different levels of current amplitude while the identity of the sensor would indicate the direction of current flow. The sensor array would then provide an array of labelled lines coding for both direction and intensity.

Inclinometers

Crustaceans perceive their orientation relative to gravity with a specialized organ, the statocyst (Figure CS 6.7), which is located in a pit at the base of the antennules (Cohen 1955). The statocyst contains two types of receptors: statolith hairs which discharge when the statocyst is at different orientations relative to gravity; and thread hairs which code for angular rotations. Statolith hairs exhibit range fractionation in the roll and pitch planes, i.e. each responds to part of the range. A biologically based implementation of a statocyst (Figure CS 6.7) might consist of a Micro-Electro-Mechanical System (MEMS) using cantilevers loaded with end masses to generate orientation-dependent signals.

Lobster behaviour

The atom of structure in the behavioural sequencer is the action component. Action components form the basic unit of transition and consist of the event type (forward, stand, etc.), the previous event at the time the action pattern is evoked, the duration of the event, the latency of the event (for developing sequences), and the intensity of the event. The intensity corresponds to the underlying parametric modulation in terms of amplitude and frequency. We employ two other modulatory components: (1) the speed for locomotory behaviour, and (2) the posture in the roll, pitch, and yaw planes. The intensity is analogous to parametric neuromodulation. It has a rise time, a fall time, and a peak amplitude. These action components constitute the instruction set of the robot.

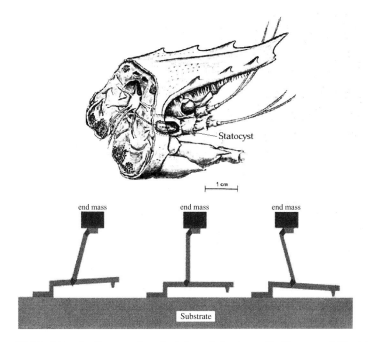

Figure CS 6.7. *A biomimetic sensor for orientation relative to gravity.* Upper panel: The lobster statocyst is located in a pit at the base of the antennules. (After Cohen 1955.) Lower panel: A biologically based inclinometer. An end mass is attached to the end of the middle cantilever of the Micro-Electro-Mechanical System (MEMS) device. Two inclinometer arrays in the roll and pitch planes would code in an analogous fashion to the crustacean statocyst.

Action components are sequenced through three other structures which correspond to the three major behavioural schemes: reflexes, modal action patterns, and goal-achieving behaviour. In reflex patterns the duration and intensity are proportional to the intensity of the stimulus. Reflexes can provide continuous modulation of evoked or ongoing behaviour. Exteroceptive reflexes respond to external stimuli and modulate command systems. Proprioceptive reflexes operate at the segmental level on individual limbs.

Modal action patterns are triggered by releaser messages from sensors in an all-or-none fashion (Figure CS 6.8a). In the controller, they consist of a list of action components which are placed in sequence in a queue. The components specify transitions of action, posture, and intensity. We are expanding this list based on empirical observations of the reverse kinematics of behaving lobsters (see above). Goal-achieving patterns have both a releaser which triggers the action pattern sequence as well as a terminator which ends the sequence (Figure CS 6.8). During execution of the action pattern the system can switch between modal action patterns based on sensor feedback. This process continues until the terminator message is received from the sensors.

150 Artificial ethology

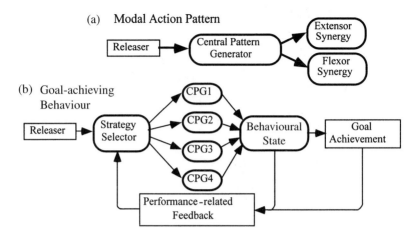

Figure CS 6.8. *Behavioural structures.* (a) Modal action pattern. (b) Goal-achieving behaviour. See text for further explanation.

An example of a modal action pattern generated by the controller is illustrated in Figure CS 6.9. In this example the releaser is collision with a wall perceived simultaneously by both antennae. The response is a sequence which consists of an increase in parametric modulation (to high speed) and a sequence of two action components: (1) several cycles of backward walking and (2) a rotation to the right or left. Where the collision is detected by only one antenna, a different action pattern consisting of a side step (several cycles of lateral walking) to the side opposite to the collision, followed by a resumption of forward walking, would allow the system to avoid the obstacle.

Finite state analysis of lobster behaviour

We have developed computer-controlled video technology for reverse animation and kinematic analysis of animal behaviour (Ayers 1992). This multi-media system allows correlated acquisition of kinematic and electrophysiological data by simultaneously recording behaviour on the video channel and electrophysiology on the audio channels of a high-resolution digital VCR. We developed extensions to a public domain image analysis program (NIH Image) which include the capability for colour-based acquisition and image segmentation as well as time-based quantification of kinematic parameters and correlated analogue acquisition (ColorImage: Ayers and Fletcher 1990; Ayers 1992). This system allows us to measure animal orientation and joint angles from video, on a frame by frame basis, to establish the detailed movement strategies and kinematics of compensatory, orientational, and taxic reflexes, along with the underlying neuromuscular control signals. As a result it has been possible to establish the co-ordination patterns and control signals underlying

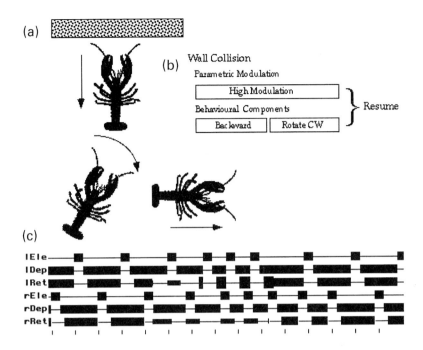

Figure CS 6.9. *Avoidance of an obstacle detected simultaneously by both antennae.* (a) The overall behavioural response. (b) The sequence of action components comprising the modal action pattern. (c) The resulting control signals generated by the ambulation controller.

omnidirectional walking (Ayers and Davis 1977) as well as undulatory swimming (Ayers 1989).

To transit directly from behaviour to robotic controls, we perform finite state analysis task groups that mediate locomotion and searching individually to determine which synergistic sets are active during different behavioural acts (Ayers *et al.* 1998).

Our analysis of the sequencing of these task groups borrows from a technique utilized by astronomers to detect motion of galactic objects. As the analysis proceeds through each frame of the digital movie, the program flashes between temporally adjacent frames of the movie with a brief pause after each cycle. Appendages that are moving the most flash clearly in these projections. A panel of buttons that represent different states of the task groups (for example elevation versus depression of the chelipeds) are available to the investigator to specify which groups are active. By clicking on the appropriate buttons for each frame, it is possible to quantify efficiently the activity of all task groups at high temporal resolution from video tapes of specimens behaving in a variety of situations. These state diagrams are used to establish control sequences for the robots based on the behaviour of the model organisms.

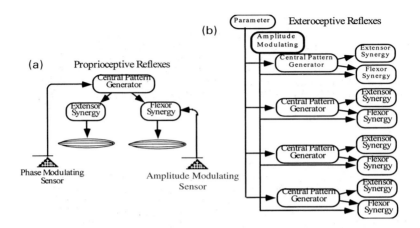

Figure CS 6.10. *Crustacean proprioceptive reflexes.* (a) Proprioceptive reflexes include phase-modulating inputs which act upon the neuronal oscillator, and amplitude-modulating reflexes which act on the motor neurones. (b) Exteroceptive reflexes including parametric modulation and amplitude modulation.

Compensatory reflexes

Lobsters and other crustaceans exhibit a variety of compensatory reflexes ranging from adaptations of movements of single limbs (Ayers and Davis 1977a, b, 1978) to multisegmental exteroceptive reflexes (Davis and Ayers 1972). Proprioceptive reflexes can be divided into two types: amplitude-modulating (acting on motor synergies), and phase-modulating (acting on the neuronal oscillators (Figure CS 6.10). Load compensation is mediated by connections from proprioceptors to motor synergies. which modulate the amplitude of propulsive synergies and delay the onset of elevator action (Pearson 1972). In the ambulation controller these messages are passed to the oscillator and recruiter components. Stumbling reflexes are triggered by stimuli that occur during the swing phase of stepping. These reflexes are mediated by prolongation of the swing phase without resetting the overall timing of stepping; their implementation has been described in detail (Ayers and Crisman 1992). Exteroceptive reflexes act on command systems and/or intersegmental modulatory interneurones. An example of an exteroceptive reflex during locomotion is rheotaxis. In this behaviour, the releaser is water movement directed from the sides, perceived by hair fan organs or their biologically based analogues (Figure CS 6.6). Where small deviations from the midline are perceived the response is a yaw correction, while larger deviations (>45°) cause an intercalated rotation to the left or right (Figure CS 6.11).

Compensatory reflex responses to water currents involve simultaneous compensation in the pitch plane as well as postural responses of the abdomen

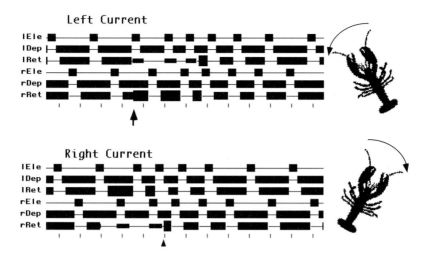

Figure CS 6.11. *Rheotaxic exteroceptive reflexes caused by water currents from the left or right.* The time of the stimulus is indicated by a vertical arrow. In both cases, the response is to rotate into the direction of the current as indicated by the diagrams at the right.

and chelipeds. The ambulation controller supports an antigravity recruiter which acts on the depressor synergy of each leg and mediates pitch and roll compensation. During medium currents the controller reduces the depression in anterior segments which will pitch the hull forward. During high currents the controller also increases depression in the caudal segments causing even greater forward pitch. These compensatory responses to water currents and surge involve both yaw and pitch plane components as well as load compensation for the necessary added propulsive thrust.

Orientation

The yaw components of orientation are mediated reactively by taxes and kineses (Loeb 1918; Braitenberg 1984). Positive yaw taxis or attraction occurs when sensor bias directs locomotion towards a source and is generally mediated by contralateral stimulation between sensors and effectors (Figure CS 6.12). Negative yaw taxis or avoidance occurs when sensor bias directs locomotion away from a source and is generally mediated by ipsilateral stimulation between sensors and effectors (Figure CS 6.12). In the ambulation controller, such inputs send messages to both the parametric command and the recruiting component. In attractive reflexes the messages are sent to the contralateral command objects while during avoidance reflexes (negative taxis) the messages are sent to the ipsilateral command objects. As a result, the controller biases the period on the two sides (producing faster walking on the side turned away from) as well as biasing the recruiters on the two sides to generate more propulsive force on the

154 *Artificial ethology*

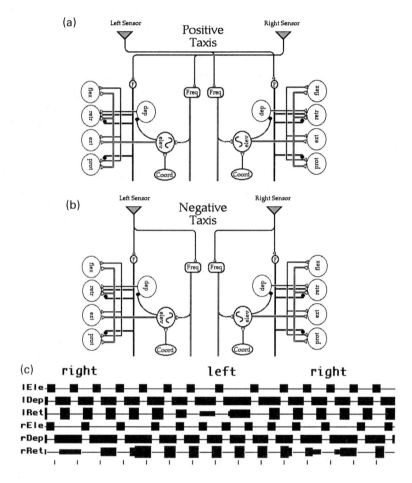

Figure CS 6.12. *Networks for yaw plane exteroceptive reflexes.* (a) Positive yaw taxis. (b) Negative yaw taxis. (c) Yaw plane modulating motor patterns during forward walking. The traces indicate the elevator, depressor, and retractor synergies of the left and right sides.

faster side. Control schemes like that of Figure CS 6.12 can be easily adapted to a variety of beacon tracking, avoidance, docking and search behaviours through acoustic, magnetic, optical, tactile, gravitational, flow, and chemical sensors.

Behavioural choice and hierarchies

In many cases, behavioural acts that operate in parallel are superimposed upon each other at the level of the effectors (von Holst 1973). In other cases, especially when presented with the releasers for two incompatible behavioural acts, the animal typically chooses to perform one act rather than the other. Animal studies have indicated some mechanisms by which such choices are made (Davis 1979). In some cases, the choice is mediated by direct synaptic interactions between the command systems that mediate the behaviour. In other cases, lateral inhibitory interactions between the releasers for one behaviour may suppress the command for another behaviour. Furthermore, circulating neuropeptides may gate off particular categories of behaviour with no effect on others (Davis *et al.* 1974). The sum of these interactions dynamically establishes a basic priority scheme or hierarchy for the autonomous behaviour of these simple animals (Davis 1979). When environmental inputs modulate the transitions between individual acts, the potential for sensor-mediated choice allows the generation of adaptive sequences of behavioural acts in response to perturbation. In fact, it has been possible to model the behavioural hierarchy of the crayfish using lateral inhibition between commands.

Conclusions

It is feasible to base the design of an autonomous underwater robot on biological principles. Sensors, controlling circuits, and actuators can readily be designed that operate on the same principles as their living analogues. Nature is conservative in the neuronal control strategies throughout the invertebrates, always relying on the command system, co-ordinating system, central pattern generator model (Stein 1978). It is worthwhile to explore in the future whether such a control architecture might also prove generalizable to different types of underwater robots.

End of Case Study 6

The three studies in this chapter show a number of different perspectives on the relationship between robots and ethology. In Case Study 4, Simon Giszter models the behaviour of the spinal frog using Maes' behaviour networks, originally developed for use in behaviour-based robots; he also uses a real robot as an experimental tool to interrogate the frog's behaviour. Holk Cruse's case study shows many influences taken from robotics, and demonstrates the power of good simulations of controlled physical systems; however, he also makes it clear that the eventual use of a robot is a necessary and not merely a desirable step. The primary focus of Joseph Ayers' case study is the development of a robot, rather than the increased understanding of the lobster, but it is clear that the pressure for what he calls 'a complete design solution' has necessarily driven his work towards obtaining a clearer understanding of the lobster as a whole animal. It is also noteworthy that he has encountered no serious incompatibility between the biologists' analysis of the control of lobster behaviour, and the information required by engineers to construct a lobster-like robot.

Chapter 5

Motivation and learning

The concept of motivation has a complex and contentious history, and many issues remain unresolved today. Plato and most of the ancient Greek philosophers regarded human behaviour as being the result of rational and voluntary processes, with individuals being free to choose whatever course of action their reason dictates. This view, called 'rationalism', persists to this day.

Thomas Aquinas, in the thirteenth century, regarded animal behaviour as being determined by sensuous desire, though he appeared to recognize some elementary process of judgement in animals: 'Others act from some kind of choice, such as irrational animals, for the sheep flies from the wolf by a kind of judgement whereby it considers it to be hurtful to itself; such a judgement is not a free one but implanted by nature'.

René Descartes (1649) held that animals were mechanical automatons, while human behaviour was under the dual influence of a mechanical body and a rational mind. A pinnacle of materialism was reached by Thomas Hobbes (1651), for whom the explanation of all things was to be found in their physical motions. For Hobbes, the will is simply an idea that people have about themselves. In accounting for mental events in material terms, in seeking mechanistic explanations of purposive behaviour, and in regarding the will as an epiphenomenon, Hobbes anticipated much modern scientific thought. However, such thoroughgoing materialism was not to become acceptable for many decades.

The associationists, like the materialists, denied any freedom of the will, but they did not necessarily attempt to account for behaviour in physical or physiological terms. Both John Locke (1700) and David Hume (1739) took the view that human behaviour develops entirely through experience, according to laws of association. This view was very influential in the early days of psychology.

Animal learning

Ivan Pavlov gave a lecture in Madrid in 1903, and he delivered the Huxley lecture in London in 1906, of which a report was printed in *Science*. A review of Pavlov's work by Yerkes and Morgulis was published in 1909, and Watson published another in 1916. A translation of Pavlov's book *Conditioned Reflexes* was published in 1927. There was a strong climate of opinion at the time in favour of a purely mechanistic and objective science of behaviour. Pavlov's work added fuel to the extreme environmentalist approach espoused by Watson in his behaviourist approach to psychology. Watson (1916) realized that reflex conditioning might serve as a paradigm for learning in general. Behaviourists such as Watson and, later, B.F. Skinner claimed that all animal and human behaviour could be accounted for in terms of conditioning. Pavlov's work endowed behaviourism with a certain amount of physiological respectability, and the psychology of animal learning developed to dominate western academic psychology up to the end of the 1950s.

The classical-conditioning paradigm

In his original conditioning experiments, Pavlov restrained a hungry dog in a harness (see Figure 5.1), and presented small portions of food at regular intervals. When he signalled the delivery of food by preceding it with an external stimulus like the sound of a bell, the behaviour of the dog towards the stimulus gradually changed. The animal began by orienting towards the bell,

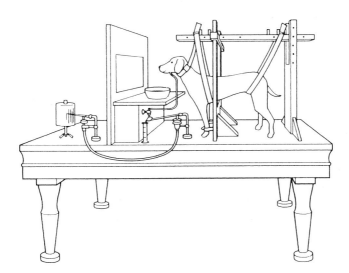

Figure 5.1. *Pavlov's arrangement for the study of salivary conditioning.*

licking its lips, and salivating. When Pavlov recorded the salivation systematically by placing a small tube in the salivary duct and collecting the saliva, he found that the amount of saliva collected increased as the animal experienced more pairings between the sound of the bell and food presentation. It appeared that the dog had learned to associate the bell with the food.

Pavlov referred to the bell as the conditional stimulus (CS) and to the food as the unconditional stimulus (UCS). Salivation in response to presentation of food was called the unconditional response (UCR), while salivation in response to the bell was called the conditional response (CR). The rationale behind this terminology is that the food unconditionally elicits a set of consummatory responses, one of which is recorded by the experimenter and designated the unconditional response. Conditioning occurs as a result of a contingency, arranged by the experimenter, between the unconditional stimulus (food) and an external stimulus previously unconnected with the feeding situation, like the sound of a bell. After a number of pairings, the bell alone is sufficient to elicit salivation. It is then known as the conditional stimulus because as a result of its training, the dog salivates if and when the stimulus is presented. Similarly, the salivation in response to the bell is known as the conditional response even though it may appear to be the same response as the unconditional response. During the process of conditioning, the presentation of the UCS (food) following the CS (bell) is said to reinforce the conditional reflex of salivation to the CS. The UCS, therefore, is regarded as a reinforcer.

A reinforcer is characterized not so much by its intrinsic properties as a stimulus but by its motivational significance to the animal. Thus, food acts as a positive reinforcer only if the dog is hungry, and an air puff acts as a negative reinforcer only if it is noxious or unpleasant for the animal. In many cases the reinforcer is innate in the sense that its motivational significance and ability to support conditioning are integral parts of the animal's normal makeup. However, this does not have to be the case, and Pavlov showed that a CS could act as a reinforcer. For example, if a bell is established as a CS by the normal conditioning procedure, it will reliably elicit a CR, like salivation. If a second CS, such as a light, is then paired repeatedly with the bell, in the absence of food, the animal will come to give the CR in response to the light alone, even though food has never been associated directly with the light. This procedure is known as second-order conditioning.

Pavlovian, or classical, conditioning is very widespread in the animal kingdom, and it pervades every aspect of life in higher animals including humans. Pavlov demonstrated that conditioning could occur in monkeys and in mice, and claims have been made for a wide variety of invertebrate animals. In evaluating such claims, however, we must take care to distinguish true classical conditioning from other forms of learning and quasi-learning.

Although classical-conditioning procedures are relatively straightforward, the phenomena they reveal are not so clear-cut and have given rise to considerable discussion in the psychological literature, from the time of Pavlov to the present day. Every student of animal behaviour should be familiar with the basic characteristics of classical conditioning because it is hardly possible to carry out an experiment without some conditioning occurring. It may be simply that the animal becomes conditioned to the time of day at which the experimenter appears or that clandestine conditioning phenomena subtly invalidate the experimenter's conclusions. In any event, as a universal feature of higher animals, conditioning is not only of practical importance but also must be incorporated into any coherent understanding of the way animals work. For a fuller elementary account of classical conditioning, see McFarland (1999).

Instinct

Early writers regarded instinct as the natural origin of the biologically important motives. Thus, Thomas Aquinas wrote that animal judgement is not free but implanted by nature. Descartes regarded instinct as the source of the forces that govern behaviour, being designed by God in such a way as to make the behaviour adaptable. The associationists appeared to reject all notions of instinct, although Locke did write of 'an uneasiness of the mind for want of some absent good...God has put into man the uneasiness of hunger and thirst, and other natural desires...to move and determine their wills for the preservation of themselves and the continuation of the species.'

While the associationists believed that human behaviour is maintained by the knowledge of and desire for the consequences of behaviour, others like Hutcheson (1728) argued that instinct produces action prior to any thought of the consequences. Whereas instinct previously had been regarded as the source of motivational forces, Hutcheson made instinct the force itself. This concept of instinct was seized upon by the new rationalists such as Reid (1785), Hamilton (1858), and James (1890) as a convenient vehicle for the non-rational elements of behaviour. Thus, human nature was seen as a combination of blind instinct and rational thought.

The idea of instinct as a prime mover was taken up by psychologists such as Freud (1915) and McDougall (1908). Sigmund Freud developed a motivational theory of neuroses and psychoses, emphasizing the irrational forces in human nature. He saw behaviour as the outcome of two basic energies: a life force underlying life-maintaining and life-continuing human activities and a death force underlying aggressive and destructive human activities. Freud thought of the life and death forces as instincts, the energy of which was seen as requiring expression or discharge. McDougall thought of the instincts as irrational and compelling sources of conduct that oriented the organism towards its goals. He postulated a number of instincts, most of which had a corresponding emotion – for example, flight and the emotion of fear, repulsion and the emotion of disgust, curiosity and the emotion of wonder, and pugnacity and the emotion of anger.

These various conceptions of instinct were derived from the subjective human emotional experiences. This essentially unscientific practice involves difficulties of interpretation, of agreement among different psychologists, and of determining the number of instincts that should be allowed or recognized. Darwin (1859) was the first to propose an objective definition of instinct in terms of animal behaviour. He treated instincts as complex reflexes that were made up of units that were compatible with the mechanisms of inheritance. Instincts were thus the product of natural selection and had evolved together with the other aspects of the animal's life. Darwin's concept of instinct is thus similar to that of Descartes, with evolution replacing the role of God.

Darwin laid the foundations of the classical ethological view propounded by Lorenz and Tinbergen. Lorenz (1937) maintained that much of animal behaviour was made up of a number of fixed-action patterns that were characteristic of the species and largely genetically determined. He subsequently postulated that each fixed-action pattern, or instinct, was motivated by action-specific energy (Lorenz 1950). This was likened to a liquid in a reservoir. Each instinct corresponded to a separate reservoir, and when an appropriate releasing stimulus was presented, the liquid was discharged in the form of an instinctive drive that gave rise to the appropriate behaviour. Tinbergen (1951) proposed that the reservoirs, or instinct centres, were arranged in a hierarchy so that the energy responsible for one type of activity, like reproduction, would drive a number of subordinate activities, such as nest-building, courtship, and parental behaviour. Lorenz and Tinbergen provided numerous examples of what they regarded as instinctive behaviour patterns.

The innate releasing mechanism

The early ethologists thought that animals sometimes respond in an instinctive manner to specific, though often complex, stimuli. Such stimuli came to be called 'sign stimuli', an example of which is illustrated in Figure 5.2. A sign stimulus is part of a stimulus configuration, and it may be a relatively simple part. For instance, a male three-spined stickleback (*Gasterosteus aculeatus*) has a characteristic red belly when in breeding condition. This is a sign stimulus that elicits aggression in other territorial males. As we can see from Figure 5.2, crude models suffice to elicit aggression provided they have a red underside. In contrast, a freshly killed male stickleback without a red belly is ineffective in provoking attack from other males. Thus, many of the details of the structure and texture of a male stickleback are apparently ignored by other males. The red coloration is much more effective if it is on the underside of the model.

Such a configurational relationship is a common feature of sign stimuli. In the case of sticklebacks it does not seem to be an essential feature, however.

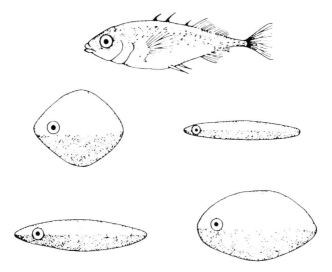

Figure 5.2. *Models of male sticklebacks that elicit attack by territorial males.* At the top is a dead male lacking nuptial colours. The four lower models are very crude, but their red underside provides a sufficient sign stimulus to elicit attack. (After Tinbergen 1951.)

Tinbergen (1953) describes how he was studying the behaviour of territorial male sticklebacks in aquaria placed in a window. Whenever a red post office van passed along the road outside the window, the sticklebacks immediately attempted to attack it as if it were a rival male. The BBC was able to repeat these observations while making a historical documentary film for television (Sparks 1982).

The selective responses to stimuli suggested to early ethologists that there must be some built-in mechanism by which such sign stimuli were recognized. This supposed mechanism came to be called the 'innate releasing mechanism' or IRM (Lorenz 1950; Tinbergen 1950). There are three important aspects of this concept. Firstly, the mechanism is envisaged as being innate in the sense that both the recognition of the sign stimulus and the resulting response to it are inborn and characteristic of the species. (Note, however, that the notion of innateness used by early ethologists is somewhat different to that current today.) Secondly, the IRM has the role of releasing the response to the sign stimulus. The implication here is that the IRM holds back the pent-up action-specific energy, or 'drive', until the appropriate sign stimulus is recognized. The energy is then released in the form of appropriate behaviour. So central was this aspect of the IRM that sign stimuli were referred to often as releasers. Thirdly, the response released by the IRM was stereotyped and was part of the animal's innate repertoire of fixed-action patterns. Fixed-action patterns, as originally conceived by Lorenz (1932), were activities with a relatively fixed pattern of co-ordination, somewhat akin to reflexes. They were innate and typical of the species.

In more recent times, it is recognized that many typical fixed-action patterns are modifiable by learning, as we see in the next case study.

Case study 7: Neural nets and robots based upon classical ethology   *by Janet Halperin*

The Siamese fighting fish (*Betta splendens*) is a good vehicle for illustrating the tenets of classical ethology. It has distinctive fixed-action patterns, and apparent changes in mood, or background motivation. Its behaviour can be readily studied in the laboratory, and it is a good subject for learning experiments.

Studies of the behaviour of this fish have inspired neural modelling based directly upon observation of behaviour, rather than upon abstract theory, the more common approach. These neural models have, in turn, inspired robot control systems, as we see in this case study.

Fixed-action patterns and motivation in animals

The classical ethologists saw that an animal's stream of activity can be broken up into quantifiable 'fixed-action patterns', recognizable, repeated patterns within what at first appears to be a continuous flow of activity. Fixed-action patterns were thought of as being instinctive units of behaviour, passed down genetically from one generation to the next, and thus recognizably specific to the species.

The aggressive behaviour of Siamese fighting fish is our exemplar. When Siamese fighting fish begin to fight, they first perform various threat postures, or displays (see Figure CS 7.1)

A fighting fish's major threat postures are 'facing' display (swinging to face a rival while raising the gill covers and lowering the branchiostegal membrane to form a large dark halo below the head), and 'broadside' display (swinging broadside to a rival, slightly lowering the gills but spreading the fins) (Simpson 1968). Later the fish may 'tailbeat', pushing water towards the opponent. Then come contact behaviours such as mouth-fighting (mutual grab and shake) and bites directed at the opponent's large fins.

The main function of these displays is communication. For example, the intensity of a threat display communicates a fish's strength and motivation towards opponents competing for space, food, or mates.

Figure CS 7.1. *The displays of the Siamese fighting fish* Betta splendens.

Fixed-action patterns have six main properties:

1. The first property of fixed-action patterns is that they are relatively discrete and recognizable. The aggressive displays of animals are often identifiable by a stereotyped orientation as much as by a stereotyped body posture. In Siamese fighting fish threat, orientation is the defining component, hence the names 'facing' and 'broadside' (Figure CS 7.1). On reflection, this is not surprising. Social behaviours communicate to a recipient, and must be oriented in a particular way to be received as a stereotyped signal. The significance of this for those modelling fixed-action patterns lies in the difficulty of specifying orientations as sequences of muscle contractions. Fixed-action patterns are in fact succinctly describable by saying that they achieve a perception, whether proprioceptive or exteroceptive. In *Betta* the fixed-action pattern must include a specification of the visual input associated with being broadside to the object of fixation (the opponent fish, or a mirror used in an experimental study).

2. Fixed-action patterns are less stimulus-bound than reflexes. Their intensity is not a simple function of the stimulus intensity, and their timing is not so closely determined by the stimulus. The threat displays of *Betta* persist after the

stimulus disappears. If a displaying fish suddenly has no opponent, perhaps because it has been removed by the experimenter, the fish's display will persist, a phenomenon called 'after-discharge'. After-discharge occurs on two timescales. The discrete fixed-action patterns' after-discharge lasts for a second or two, but there is a much longer aggressive after-discharge with an alternation of threats (see below).

3. Fixed-action patterns can interrupt each other, and in the normal course of events they do so. A Siamese fighting fish that is eating will stop eating and display if a rival appears. This property has obvious adaptive value, and a fish would be very inflexible if behaviours switched on for a fixed time regardless of changes in the stimulus situation.

4. Classically, fixed-action patterns are released by 'sign stimuli', usually particular components of the stimulus provided by a rival, a potential mate, etc. These components may summate in their effect upon the recipient, a phenomenon known as 'heterogeneous summation'. For example, the stimulus provided by a rival *Betta* contains many components that are part of the total sign stimulus. The colour is one feature that helps to release aggression. A green (model) rival is not as effective as a red one. Texture is important, with a fish-scale texture more effective than a smooth one. Shape is relevant, with fin-like angled shapes more effective than smooth ovals. Deviations from the optimal key stimulus in any dimension smoothly and gradually reduce the effectiveness of the stimulus configuration.

5. Fixed-action patterns were originally thought of as being fixed in the sense of innate, but they are now known to be modifiable by learning. Fighting fish quickly learn in which stimulus situations to perform their threat displays. Different opponents that provide the same sign stimuli elicit quite different responses. With experience, the fish will come to flee from one, attack another, and ignore a third, indicating learned individual recognition. Similarly, the fish learn the appropriate level of aggression for their resource-holding potential (Halperin *et al.* 1997).

6. The fighting behaviour of fish is accompanied by physiological arousal, which persists after the rival disappears. There is a relatively long after-discharge period in which an aggressively aroused fish will go on performing sequences of threat displays even if there are no more aggression-eliciting stimuli present. During a fight, the fish's colour darkens, and this darker colour is usually retained for a long period after the fight.

There is a lowered threshold for performing specific fixed-action patterns long after overt, spontaneous display has faded. This presumably reflects the persistence of the state even when overt aggressive behaviours have stopped being expressed. However, the motivational arousal leaves the decision about exactly which behaviour to perform open, to be influenced by the moment-to-moment stimulus situation.

Each motivational state biases the animal towards performing a certain group of fixed-action patterns, the subset of potentiated behaviours defining the particular state. In this sense the motivational states are discrete, and delineated by a particular subset of potentiated behaviours. Some behavioural components may be common to many motivational states, especially simple approach or avoidance patterns.

Ideally, an animal-like robot would show fixed-action patterns and motivation with all the properties outlined above.

Implementing fixed-action patterns by machine motivation

Machine motivation is simultaneously a robot adaptive control system model, which allows robots to have animal-like fixed-action patterns and motivation, and a working hypothesis about the neural principles underlying animal fixed-action patterns and motivational states. We have begun testing it as an ethological model, through its predictions of new phenomena in fighting fish aggression. In this section, we shall outline how the machine motivation system implements the properties of fixed-action patterns described above. A typical module is sketched in Figure CS 7.2.

Fixed-action patterns are discrete. This property is captured naturally because the system is inherently modular. One motivation module is assigned for each fixed-action pattern.

Each module consists of a pool of releaser neurones (R) and a pool of behaviour command neurones (B), functionally separate from the R–B pools of other modules. The module's output, being a pattern of activation in a neurone pool, could activate any suitable mechanism for achieving a fixed-action pattern.

The temporal properties are achieved through feedback, some positive and some slower-acting and negative. Temporal persistence is achieved in any neurone through facilitation, a within-neurone positive feedback process, but in machine motivation stronger persistence can be implemented by positive feedback between the B and R neurone pools. This allows after-discharge. Again, the specificity of the feedback from the B pool to the module's R pool defines the module. Temporal limitation happens by neural adaptation, a spontaneous increase in threshold occurring when neurones are active (or for perceptual functions, when they are receiving input). This is a slower-acting negative feedback process.

Competitive inhibition amongst modules is implemented by inhibitory connections amongst B pools. Initial biases in the strength and numbers of S–R

166 Artificial ethology

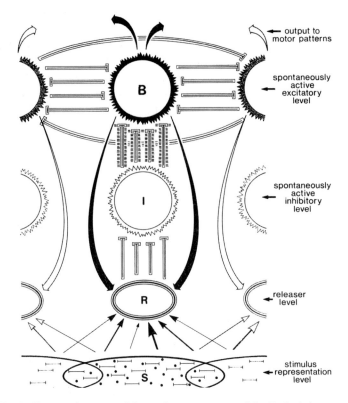

Figure CS 7.2. *The type of wiring used for machine motivation modules.* Each circle represents a pool of neurones. Arrows are excitatory one-way connections. Blunt-ended arrows are one-way inhibitory connections. The connections from S to R change strength according to the neuro-connector rules. There are local inhibitory relationships within the S pool, and there may be within the R and B pools. These could gate neighbours or inputs to neighbours. This version releases behaviours by inhibiting I units, which normally inhibit spontaneously active B units.

connections can be introduced at the design stage. The connections are set in animals by evolution, and in robots by an initial phase of 'demonstration design' (Hallam *et al.* 1994).

The capacity for heterogeneous summation is implemented by using biologically plausible neuro-synaptic properties. These include overlapping receptive fields and 'fan-in fan-out' local connectivity. Finally, learning is achieved by allowing the neural network to change excitatory connection strengths after the initial demonstration design process. It must also be ensured that generalization is maintained when learning starts to modify the connection strengths, so it does not use the popular algorithm that minimizes 'least mean squares', but a biologically inspired algorithm (NNAC – Neural Net Adaptive Controller) which uses only information local to the connections and which guarantees generalization even when the training set is not an 'independent identically distributed' random sample.

During learning, as the S–R (stimulus–response) connections change strength to allow new stimulus situations to trigger fixed-action patterns, the connections must incorporate information about whether a particular fixed-action pattern was successful in a particular situation. This requires modelling how local firing pattern information alone provides enough information for individual synapses to determine whether a given behaviour was successful.

Implementing motivational states

The machine motivation model implements motivational states, called emotional states by Halperin (1995), in much the same way that it implements discrete, communicative fixed-action patterns. The early ethologists observed that the simpler low-level fixed-action patterns and the more complex states had many commonalities. The machine motivation modules inherently generate the features that motivational states have in common with fixed-action patterns: release by sign stimuli, persistence of activation, competition amongst states, and learning processes that use feedback about success to refine the assessment of the appropriate situations for showing the state.

We can obtain the important difference in temporal persistence between motivational states and discrete fixed-action patterns by increasing the time constants for adaptation. The output signal from the B pool of a single state module can go to a variety of places. Motivation modules for states can trigger autonomic responses by direct activation. They can also potentiate approach, avoidance, and selected groups of fixed-action patterns when state modules are arranged in a separate layer and send excitatory signals to the motivation modules for fixed-action patterns. Factors down-line from the state modules will affect whether excitatory signals from the state modules would actually activate a fixed-action pattern (for example, competitive inhibition amongst fixed-action patterns), so the state module will potentiate rather than command fixed-action patterns.

Partial competition between state modules may be accomplished by setting the inhibition between B modules very low, or by local lateral inhibition within each small subregion. This local lateral inhibition is also assumed in the S pool, where it would implement attention (Figure CS 7.2) by limiting the number of neurones active in a region.

To illustrate why this is useful, consider a small subregion with neurones active in response to various inputs. As the neurones in the subregion compete for activation, there is a local decision about what motivational state is assessed as most appropriate on the basis of the local input factors.

The outcome of the competition can be different in different subregions, allowing the activation of mixed motivational states combined with considerable competition amongst states.

Thus the machine motivation modules can be used without significant modification to model the longer duration states of motivational arousal, which bias behaviour selection towards the pre-selected set of relevant fixed-action patterns, and elicit the machine equivalent of autonomic responses preparatory for actions.

Implementing learning about the success of behaviour

We now want to model how a Siamese fighting fish learns very quickly that a carefully made, realistic physical model of a fighting fish is not the same as a live opponent, because it does not respond to aggressive display in the same way. Such learning about behaviour outcome is an inherent part of the machine motivation model, occurring within the architecture, which organizes motivational states and fixed-action patterns. The S–R (stimulus–response) connections self-calibrate on the basis of feedback about the success of the behaviour.

Our model is connectionist, and forms associations on the basis of local information in pairs of connected neurones, so there must be a process in the connections themselves to 'decide' if a behaviour was successful. The neuro-connector rule governing learning about the success of behaviour is a dynamic, analogue relative of Hebb's rule (Hebb 1949). Hebbian processes are regarded as simple, but biologically plausible. Each neuro-connector keeps decaying traces of its own recent inputs and activity. The activity trace controls connection strength changes and can interfere with consolidation of strength changes.

The model is unusual in several ways. It associates firing bouts, rather than associating neural activity at each moment of time as do most such connectionist systems. The neuro-connector system 'decides' that a firing bout has occurred when there is neural activity followed by silence, in effect looking for temporal edges in firing patterns. It strengthens the association between two neurones if the temporal edges in their firing patterns follow one another at or close to a specified time interval τ (Figure CS 7.3). In this way the system associates neural events that follow each other at a certain time interval. An S–R connection weakens if the temporal edge in R is absent, ends too early, or ends too late (Figure CS 7.4).

Factors that determine the temporal pattern of firing in neurones are crucial to whether a connection forms between two neurones, so we shall consider what controls the firing patterns of the neurones that implement fixed-action patterns in Figure CS 7.2, starting with the S neurones. These fire rather straightforwardly in response to stimulus situations, responding either to the

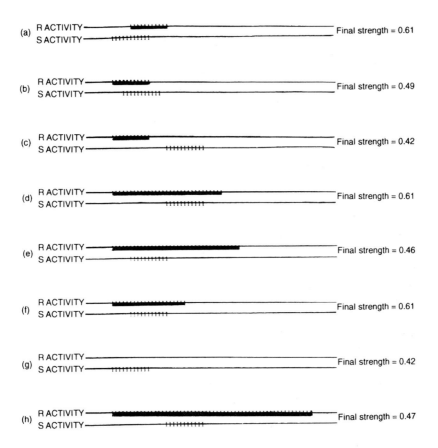

Figure CS 7.3. *The neuro-connector rule in action.* Examples of how various patterns of firing in a presynaptic neurone (S) and its postsynaptic neurone (R) affect the strength of the connection between them. (a) Time-lagged Hebb-like strengthening. (b) and (c) Weakening of association due to a truly backward temporal relationship. (d) Strengthening because S is active just before the end of firing in R. (e) Weakening because the S bout stops too early with respect to the end of R's activity. (f) Strengthening as in (a) and (d), showing that the relative timings of the beginnings of activity bursts have no effect. (g) An S bout not followed by firing in R should weaken any existing S-R connections. (h) This illustrates that the lengths of bouts are not critical (assuming that a minimum duration to 'fill' the activity trace is achieved) and also that the exact relative timing of endings producing weakening is not very precise – compare with (e). (From Halperin and Dunham 1992.)

onset of a stimulus factor or to the presence of some stimulus factor. We assume that many S neurones respond not to stimuli per se, but to stimulus onsets, as in animal nervous systems. This is modelled by having many S neurones adapt quickly (they fire at the onset of stimulation, but then stop firing whether or not the stimulus continues).

The R neurone timing, however, is not controlled so directly by the stimuli, but instead by the success of the behaviour that the R neurone releases. The R

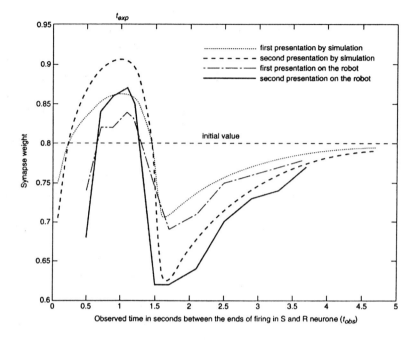

Figure CS 7.4. *The temporal characteristics of the neuro-connector rule.* This figure shows how the strength change in a connection depends on the time interval between the temporal edge in the presynaptic (S) neurone and the temporal edge in the postsynaptic (R) neurone. The x axis is the time until the R edge after the S edge has occurred. The nearer the timing is to the optimal value τ, the better will be the strengthening. The data are from successive presentations of stimuli to a tethered robot and from similar simulation runs. (As described in Hallam 1993.)

neurones fire in response to a stimulus situation, since they are activated by S neurones, but they fire persistently because of the R–B feedback loop. The all-important end of firing, the temporal edge, is controlled by what happens next. If the animal is involved in a biologically significant interaction, what happens after the behaviour is released is likely to be that another, different behaviour is released. If a new behaviour is released (and this is our indicator of success) the competitive inhibition amongst B units allows the newly activating B neurone pool to break the original R–B loop. If the new behaviour was released at around τ seconds after the end of the firing bout in the S neurone, the activity timing profile will strengthen the S–R connections (Figure CS 7.3a). Thus the triggering of a subsequent behaviour at about the specified time after an S–R sequence catalyses the strengthening of this S–R pathway. If there is no biologically significant consequence, no other fixed-action pattern is triggered, the R–B loop continues much longer, and the activity pattern is that in Figure CS 7.3e. The connection weakens from those S neurones representing the onset of the stimulus situation. Any S neurones that responded to the situation continuously (or fired only near the end of the R–B loop after-discharge) will have a temporal profile of bout endings like that in Figure CS 7.3b or c (depending on how quickly the fish turned its attention from the stimulus once the after-discharge ended). Either profile will weaken these S–R pathways.

The logic is that the release of a subsequent behaviour indicates that the performance of an S–R sequence was successful, and that a behaviour worth reinforcement is one that altered the environment in such a way that a new response became possible. Consider an example from fighting fish. If a fighting fish performs a facing display to a rival fighting fish there will soon be a counter-display of facing posture. This counter-display will trigger a switch to broadside posture within a few seconds. This succession of behaviour and consequent behaviour allows the pathways to strengthen, as described above. On the other hand if the fish performs a facing display to a leaf there will, of course, be no counter-display. The lack of biologically significant consequences means that no subsequent behaviour will be released and the pathway will weaken.

This arrangement allows learning about the success of fixed-action patterns, which have rather immediate consequences that trigger the release of another fixed-action pattern. Learning involving motivational states cannot depend on the rapid arrival of consequences. Their consequences occur many minutes after the state is aroused (after poisoning even hours later). This is no problem, because time constants are explicitly settable in the motivation modules, so the time interval τ can be made long enough that late-arriving consequences will strengthen the S–R connections. The deviation permissible in the timing of the consequence must be increased also. This will cause more context conditioning for states than for discrete behaviours, but it might be advantageous for states to be aroused by less specific triggers than are the individual fixed-action patterns.

Testing the machine motivation system on robots

If we really understand the neural principles that mesh instinct and learning in animals, we should be able to recreate them in robots. The machine motivation system predicts some new and ethologically surprising phenomena, which have been experimentally verified in the aggressive behaviour system of Siamese fighting fish (Halperin and Dunham 1992; Halperin *et al.* 1992) But one of the most exciting things about the machine motivation model, beyond its ability to account for animal behaviour, is that it is so explicit that it can be implemented on robots.

Although the full machine motivation model (Halperin 1991, 1995) has never been implemented on a robot, two projects have implemented elements of the model. The 'TRex' project focused on mathematically analysing and implementing an important aspect of the 'neuro-connector' update rule (Johnson *et al.* 1997), and a project using a small robot called 'Ben Hope' was a first step towards a more complete implementation (Hallam *et al.* 1994). Ben tested a neurobiologically inspired analogue version of the update rule, and began the process of testing how the update rule and the architecture interact to allow an animal or robot to recognize reinforcement autonomously, without the designer having to define specific positive and negative reinforcers. The bigger, more computationally expensive challenges of using machine motivation for hierarchical autonomous learning control and for perception have not yet been tackled.

The Ben Hope project

Ben Hope, a knee-high, cylindrical, autonomous robot on wheels (Figure CS 7.5), was used for the first steps of testing the machine motivation model. The robot had been developed at the Department of Artificial Intelligence at the University of Edinburgh, where the project was carried out. It implemented a subset of the complex machine motivation model, concentrating on a type of S–R learning, and not S–S learning. This project had two goals, to test a particular analogue neural implementation of machine motivation, and to test whether the timing of reinforcement in a real laboratory situation was sufficiently accurate to strengthen the machine motivation's connections suitably.

Physically, Ben's wheeled base had an eight-segment bumper, with a contact sensor on each segment. For the tests it also used six active infra-red (IR) units. Two were mounted near the base and pointed forward, two upper IR sensors were mounted above them, near the top of the robot, and two upward-facing IR sensors were mounted on the top of the robot. Each pair had a sensitive IR unit set to detect white surfaces at about 60 cm (far detector), with the partner IR unit set to detect white surfaces only at about 20 cm (near detector). It could thus distinguish whether there was nothing in front of it, something in the mid distance, or something close in front of it. It could detect whether there was something in contact at eight locations around it, whether there was something above at about the distance of a table, or if it was being touched on the top. The computer program ran on one of Ben's four Transputer boards.

The neural net used 21 sensor neurones, some representing combinations of sensor activations. (In this very simplified implementation, there was no explicit overlap in the receptive fields, and no interneurone layers (hidden layers) to allow associations amongst sensory representations as described by Halperin [1995].) These sensor neurones were fully connected to four releaser neurones, each of which activated one of the four behaviour neurones.

The four behaviours were pre-programmed using non-neural code, although ideally a neural implementation would be used and the single behaviour neurone would then be replaced by a pool of neurones. The programmed behaviours were advance, avoid, wiggle, and afterpat (a back-up-and-turn movement). Ben had no monitors of its internal state, such as battery power, at this stage.

Figure CS 7.5. *The experimental robot Ben Hope.*

In the machine motivation model each behaviour-releaser complex has a positive feedback process within it, so that once a behaviour is triggered it persists even if the releasing stimulus disappears, a phenomenon called after-discharge (see above). This was implemented on Ben by having each R–B neurone pair activate each other. Neural adaptation in the behaviour neurones sets an upper bound on the length of the behaviour bout, so if the releaser disappears and nothing interrupts the animal the behaviour will eventually stop. In Ben this was implemented by a time-out, for simplicity.

In animals, performing a behaviour appropriately creates a new stimulus situation, which releases a new behaviour. This new behaviour may inhibit the initial behaviour. In Ben this inhibition came from any competing behaviour, as all behaviours within a level inhibited each other. In this way the timing of the new behaviour controlled the timing of the ending of the initial behaviour.

The Ben project was a first simple test of the hypothesis that reinforced learning can be implemented by installing a crude template, and using timing to refine it. The machine motivation model hypothesizes that the timing of a subsequent behaviour, detected neurally by the ending of the initial behaviour, determines whether a recently active sensor situation develops a stronger connection to the initial behaviour (Halperin 1991, 1995). This hypothesis avoids the usual over-simplifying assumption of neural net models that some situations are inherently positively reinforcing and others are punishing, and that they have opposite effects on connection strengths.

The reinforcer was a 'pat on the head', detected by the upward pointing IR pair on the top of the robot. The behaviour being reinforced was wiggling in front of objects. The neural net was set up with a crude template of the situations in which it should wiggle, and learning involved refining these situations. Ben initially wiggled whenever something activated the 'something near and low' IR sensor simultaneously with either of the upper forward-facing IRs. He was 'patted on the head' for wiggling in front of garbage cans along the wall, which activated the feature sensor neurone responding to 'something near and low plus something in mid-range up high'. When he wiggled for a blank wall, the feature detector for 'something near and low plus something near and high' was activated, and the wiggle behaviour was not rewarded with a pat on the head. Being patted activated the behaviour 'afterpat', which inhibited the wiggling behaviour. If Ben was not patted there was no other behaviour activated, and the wiggling behaviour after-discharged until it timed out. The time-out occurred at an interval that was intended to weaken connections to wiggling.

The result, as hoped, was that Ben kept wiggling for garbage cans, and stopped wiggling for blank walls. Ben also showed postponed conditioning, and weakening of connections when an action was not followed by 'biologically significant' consequences (a pat on the head that triggered another behaviour, afterpat). One may wonder why this is at all surprising, since he was in a sense being 'programmed' to be reinforced by patting, but the issue was the complexity of the machine motivation program. There were two main aspects of this complexity. Firstly, the machine motivation neurones and 'synaptic' updating were implemented using components that could be realized in chemical systems, components that were mostly differential equations. Secondly, there was an issue about having to set the values of dozens of parameters. The decay constants for the differential equations, for example, have parameters that have to be set before runtime.

Testing an analogue version of machine motivation was important because it is intended as a hypothesis about how animal nervous systems could operate. Animal nervous systems use chemicals, not computer memory arrays, so we wanted to simulate machine motivation using analogue processes that could store information about the recent past in chemical activity (although this was of course all simulated in computer memory arrays). For example, machine motivation neurones have 'chemical' traces which hold information about whether they have recently been getting inputs, and information about whether they have recently been firing. The decisions of synapses to strengthen are based on the ratios of 'chemical' traces of activity in the connected neurones.

The analogue processes all used the same simple differential equation, such that each process built up as a negative inverse exponential towards an asymptote when it was being added to, and decayed exponentially towards its floor when it was not. Thus, when a neurone was active its 'activity trace' would build up rapidly, and when it stopped firing the trace would fade at a predictable rate. Given that the neurone's activity had taken the 'activity trace' near asymptote, the amount of activity trace left can tell the system when the bout of firing ended. This is, in effect, a model of how the neurone can read out information about how long ago something happened. This type of timing information is needed in machine motivation, since the relative timing of events controls reinforcement, so we wanted to test whether such an implementation could provide it.

There is also an issue concerning the number of parameters. This implementation had many parameters. However, they were almost all related to timing, and existed because of their relationship to timing, and so relatively few of them were actually 'free' parameters. They were constrained by the kinds of behaviours the robot did and how long it would take to do them, the time it would take before consequences like a pat on the head arrived, and so on. Nevertheless there was a question of whether the set of parameters that worked to give each neurone and neuro-connector the properties needed to generate behaviour would all work together to give the ensemble of desired behaviours and desired learning phenomena.

Finding a coherent set of parameters proved not only to be possible, but easy. The neural net designer observed what kind of behaviours the robot did, how long they took, and roughly what sensor situations they should be performed in, from seeing the behaviours that had been pre-programmed. This information was sufficient to constrain most of the parameter choices. It took a few tries to realize that the rise-time parameter for the activity trace registers had to be quite large, so the activity register went to near asymptote very quickly once it had started a bout of activity. (The neurone model biases neurones to fire in bouts.) Once the need for setting the parameter high was realized, it was clear that this applied to all neurones involved in learning, because otherwise their 'synapses' could not use the amount left in the activity trace to read off time since the firing bout ended.

Ben was a successful, if very preliminary, test of these aspects of the machine motivation model. Details of the experiments can be found in Hallam *et al.* (1994).

Implications for animal behaviour

The model of reinforcement that was tested on Ben Hope allows machine motivation to 'explain' many aspects of reinforcement in natural situations (Halperin 1991). It can explain, for example, how eating can sometimes reinforce wheel running in laboratory rats, and wheel running can sometimes reinforce eating. It can explain how punishing aggression can increase aggressiveness.

As well as providing explanations for a diverse set of phenomena which must otherwise be explained separately, the machine motivation model of reinforcement predicts a novel reinforcement phenomenon, called postponed conditioning, in which a neutral stimulus presented just before a behaviour spontaneously ends comes to elicit this behaviour on future occasions. This phenomenon was observed in Siamese fighting fish. A neutral stimulus was presented just as an after-discharge of aggression started to fade out spontaneously, and the neutral stimulus itself came to elicit aggression. Because the postponed conditioning of aggression is a functionally meaningless learning capacity, a mere artefact of using the endings of neural firing bouts to signal reinforcement, finding it provided strong support for the model.

Normally, a competing behaviour would have to occur quite quickly to strengthen a connection, and machine motivation assumes that biological systems rely on the unlikelihood that an inappropriately produced fixed-action pattern would be followed quickly by an opportunity to perform another fixed-action pattern, and so be accidentally reinforced. There is an obvious exception where an inappropriately produced fixed-action pattern would usually be followed quickly by another fixed-action pattern, namely reaction to punishment. In machine motivation, properly timed punishment would positively reinforce a connection, and surprisingly, this turns out to be biologically plausible. The suppressive effects of punishment nevertheless appear in machine motivation, because avoidance tendencies are also reinforced. This would happen reliably because punishing situations elicit properly timed avoidance behaviours. The stimulus situation would develop strong tendencies to elicit avoidance and to elicit the original fixed-action pattern, and avoidance would normally dominate and inhibit the punished fixed-action pattern. This is actually consistent with a classical theory of punishment, and with the suggestion that punishing aggression makes animals both hyper-aggressive and hyper-fearful.

The machine motivation model hypothesizes that proper timing of the end of a behaviour is critical. Obviously, an animal will not learn about a consequence that arrives tomorrow. On the other hand, real-world consequences do not arrive instantaneously, and never arrive before their cause. There is a time window within which a behaviour must terminate in order for it to be reinforced. In the situation that Ben was modelling, namely a sequence of fixed-action patterns, an appropriately performed fixed-action pattern should create a situation in which another action is released within a roughly predictable time, but not at a precisely predictable time. The Ben project confirmed that the neural implementation could easily be set up with parameters that would make this window large enough that a reinforcer presented by a person in a normal, somewhat distracting laboratory situation would be accurately enough timed that the appropriate neuro-connectors would strengthen when reinforcement was given, and not otherwise.

TRex

TRex is a huge hydraulic front shovel that was retrofitted by the Lockheed Martin Advanced Environmental Systems group for remote operation in nuclear waste cleanup (Figure CS 7.6). TRex began life as a Caterpillar 235D front shovel excavator, an enormous, reliable machine whose old-fashioned open-centre hydraulic system can withstand high temperature, and dirty operating conditions. Designing a control system that would give the precision needed for picking up barrels of nuclear waste without piercing any was particularly difficult because each of the hydraulic cylinders has highly non-linear characteristics, and they are coupled because the hydraulic fluid exiting one cylinder can be the source of fluid for another cylinder. An even more serious complication for the control system designer is that the machine's response characteristics drift frequently as temperature and dirt levels vary, and because of a hydraulic phenomenon called anticavitation flow. The time lag before the system responds to a command also varies depending on a complex set of conditions. The drifting response characteristics were not a big problem in the tasks for which the Caterpillar 235D was originally designed. On construction sites it is driven by a highly trained operator, who learns to 'feel' the machine and apply more power to the hydraulic actuators if the machine's behaviour is sluggish, and less if it is responding sensitively. The Cat 235D was, quite simply, not designed for high-precision remote operation.

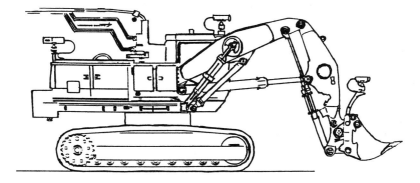

Figure CS 7.6. *A drawing of the modified Caterpillar 235D, TRex.*

A remote operator could not feel the machine without an elaborate virtual reality interface. An adaptive controller was a possible solution. This controller would detect how the machine was responding, and learn how to adjust the power accordingly. It would then obey the operator's remotely given position commands despite drifting response characteristics.

The logic of the adaptive controller was simple. The human remote operator, using visual displays and a joystick, sends position commands, indicating at each moment the desired position of the shovel tip in space. A special algorithm designed by Lockheed Martin converts this into commanded hydraulic cylinder velocities (Johnson *et al.* 1997). The adaptive controller takes the velocity command at each cylinder and reads off how much electrical power to apply to the spool from this cylinder's feedforward law, which we shall call the feedforward function, since it is just a function from commanded velocity to electric power output. If the feedforward function is correct, the shovel tip will move exactly as commanded. This is the ideal situation, having a perfect feedforward function for each cylinder so the shovel will move exactly as the operator commands. However, because the required feedforward function can drift, there is also a negative feedback process to correct errors if they are detected.

The position error signal comes from a camera on the vehicle, which detects at each moment where the shovel tip actually is, and from this calculates where the shovel tip is relative to the commanded position. If there is a position error, say the shovel tip did not move quickly enough because the machine just got more dirt in its hydraulic fluid and is being more sluggish, the adaptive controller does two things. It immediately calculates a corrected command for each cylinder so the shovel tip speeds up, and it also uses each corrected command as a teacher to adjust the feedforward function, so that the next time this command is given more power will be applied and the shovel tip will move faster. After a few iterations, the controller will have learned the new 'perfect' feedforward functions, so there will be less position error for the feedback element to correct for.

To learn the feedforward functions, a neural function approximator was a candidate, because such algorithms are well suited to learning functions iteratively from 'training' sequences. Neural systems can do this efficiently through 'generalization', which allows learning from one example to help the neural network take a better guess about what to do in similar situations.

A CMAC algorithm, often used by engineers for adaptive control, was an obvious candidate here. CMAC can be successfully used in many learning situations, and it has a biological flavour, having been originally proposed by James Albus on the basis of neurobiological analogies. It uses 'neurones' with overlapping receptive fields. For TRex, the commanded cylinder velocity would be 'place-coded' using neurones with overlapping receptive fields. With such receptive fields, each neurone is activated in proportion to how close the current commanded velocity is to its receptive field centre. A single commanded velocity activates a group of neighbouring neurones, whose activations sum to one. (You can think of a neurone's activation as its estimate of the probability that this velocity command is at the centre of its receptive field.) Each neurone also has a connection to the single output neurone, and this connection has a weight. The weight multiplies the sense neurone's activation and the result goes to the 'output neurone', which combines the signals from all its input neurones to calculate the power output. The weights collectively represent the feedforward function, as they collectively determine the mapping from commanded velocity to power output.

The logic of learning is to adjust the weights, and therefore the feedforward law, if they have commanded something other than the correct velocity output. CMAC does this by distributing the output error amongst the active neurones in proportion to their activity, and by doing this at each training step, online, it gradually minimizes the Least Mean Square (LMS) of the output error as it updates the weights. This update algorithm can converge to a set of weights that give the correct feedforward function for any input/output combinations within the training data set, but unfortunately it will not generalize sensibly under conditions that would almost certainly arise for TRex.

Instead of sensible generalization, this algorithm would give rise to weight explosion in a very ordinary situation. If the operator repeatedly commanded a set of movements that required a certain range of cylinder velocities, training the weights over and over for this range, and then suddenly went just beyond this range by commanding a movement that required a very slightly faster or slower cylinder velocity, the net would generalize terribly, and could send an enormous power surge. Naturally, this wild behaviour would be undesirable during nuclear waste cleanup, even with the remote operator a relatively safe three kilometres away and the waste site enclosed in a building.

Lockheed Martin had been investigating various other biologically inspired neural net algorithms, including Klopf's drive reinforcement algorithm, and now had a relationship with an ethological spin-off company, NeuRobotics Inc., to explore using the machine motivation system for high-level control and machine perception. They asked NeuRobotics to determine whether the machine motivation neural weight update principle, used with a single-layer receptive field type neural architecture, and stripped of its temporal dynamics, would give better generalization than CMAC. After all, animal nervous systems generalize quite well, and in machine motivation the update rule as well as the multi-layer neural architectures were biologically inspired. Stripped of temporal dynamics, the machine motivation update rule (Johnson *et al.* 1997) became sufficiently mathematically tractable that it could be mathematically proven that it generalizes sensibly, regardless of the training sequence.

The stripped-down algorithm was christened NNAC, for Neural Net Adaptive Controller, and tested using Lockheed Martin's thorough and beautifully accurate simulation of an open-centre hydraulic cylinder. Results were so positive that NNAC was put on TRex, and used in the final controller. The controller responded rapidly, improving accuracy to meet specifications easily. TRex could be intentionally programmed with an incorrect feedforward law, and quickly learned the correct function. Starting from the idea that it should be possible to strip the temporal dynamics out of the machine motivation algorithm and use it in low-level control, a mere concept, the algorithm had gone to successful application on TRex in 6 months.

This implementation demonstrated several aspects of the advantages of the NNAC. We have already discussed the advantage of guaranteed generalization, a huge advantage indeed in a situation where weight explosion could be, literally, disastrous. One of the trade-offs is that the NNAC system requires sufficient numbers of neurones, but in this case the number was still very tractable, on the order of 30. But there are other interesting points.

One issue is the reason for achieving increased accuracy with the adaptive controller, namely that it lets the machine learn a good feedforward function. Why is this better than using a good negative feedback controller?

Negative feedback control is classic and valuable, but error correction by the negative feedback controller comes with a slight delay, and this is an unavoidable problem with negative feedback. Errors have to become measurable before the calculation of correction can start, and in a system with time lags the error can have built up considerably before a correction is accomplished. The TRex system has significant time lags, and worse still, they vary, so using pure negative feedback control successfully would be very difficult.

A final set of issues involves the advantage gained by using the NNAC to make a tough open-centre hydraulic system perform with precision, rather than replacing it with a modern, high-precision hydraulic system. First, operating TRex with the NNAC demonstrated how much the feedforward function drifts, for example up to 10% over less than 20 minutes as the machine warmed up. It would have taken an enormous, time-consuming design effort to make a non-adaptive controller for TRex that could sense and correct for all the conditions that make the feedforward law drift. They are virtually impossible to model analytically. The alternative of replacing the old-fashioned open-centre system with a new high-precision hydraulic system would have been expensive, and introduces a trade-off with toughness. It would be very expensive indeed to design a high-precision hydraulic system, which could tolerate the dirt and extreme temperatures of a nuclear waste excavation site. Open-centre hydraulic systems are still used for hot, dirty jobs like mining because of their excellent toughness and low cost. The fact that they are not precision instruments is usually irrelevant for digging ore or excavating. The neural net adaptive controller was a low-cost, fast solution, which let the cheap, tough mechanism achieve precision. This is presumably a similar trade-off to the one that leads biological systems to use learning control so often.

There remain many aspects of the machine motivation model that need robotic testing. Robotic tests take a huge effort, but at least they have the potential to be conclusive. Furthermore, when we have robot-tested models of the mechanisms controlling animal behaviour, we will have expanded enormously our ability to create truly autonomous robots with well-adapted behaviour.

Acknowledgements

I wish to thank Herb Roitblat, David McFarland, Stephen Halperin, Bridget Hallam, and Ashley Lotto, who all read earlier versions of this manuscript and made helpful comments.

End of Case Study 7

Instrumental learning

While research on classical conditioning was initiated in Russia, the principles of instrumental conditioning were discovered and developed in the United States. The writings of the British psychologist Conway Lloyd Morgan (1852–1936) seem to have provided the initial impetus. Lloyd Morgan is often regarded as the father of behaviourism. In 1896, he delivered the Lowell lectures at Harvard University, where he provided the stimulus for Thorndike's pioneering research on animal intelligence. Lloyd Morgan recounted how his dog, Toby, had learned to open the latch on the garden gate by putting its head through the railings (Figure 5.3). Thorndike started his research soon after Lloyd Morgan's visit and devised ways of repeating this observation under controlled conditions in the laboratory.

Thorndike carried out a series of experiments in which cats were required to press a latch or pull a string to open a door and escape from a box to obtain food outside. The boxes were constructed with vertical slats so that the food was visible to the cat (see Figure 5.4). A hungry cat, when first placed in the box, shows a number of activities including reaching through the slats towards the food and scratching at objects within the box. Eventually the cat accidentally hits the release mechanism and escapes from the box. On subsequent trials, the cat's activity becomes concentrated progressively in the region of the release mechanism, and other activities gradually cease. Finally the cat is able to perform the correct behaviour as soon as it is placed in the box.

Thorndike (1898) designated this type of learning as 'trial, error, and accidental success'. It is nowadays called instrumental learning, the correct response being instrumental in providing access to reward. This type of learning

Figure 5.3. *A product of trial and error learning.* Morgan's dog, Toby, operating the latch of a gate. (From Morgan, C.L., *Habit and Instinct* 1896.)

had been known to circus trainers for centuries, but Thorndike first studied it systematically and developed a coherent theory of learning based upon his observations.

To explain the change in the animal's behaviour observed during learning experiments, Thorndike (1913) proposed his 'law of effect'. This stated that a response followed by a rewarding or satisfying state of affairs would increase in probability of occurrence, while a response followed by an aversive or annoying consequence would decrease the probability of occurrence. Thus, the success of instrumental learning is attributed to the fact that learned behaviour can be modified directly by its consequences. Thorndike (1911) assumed that a reinforcer increases the probability of the response upon which it is contingent because it strengthens the learned connection between the response and the prevailing stimulus situation. This became known as the 'stimulus–response theory of learning', and versions of this theory were predominant for many years. While recognizing the validity of the law of effect as an empirical statement, most present-day psychologists doubt that behaviour is modified by its consequences in the direct way that Thorndike and his followers supposed. To understand this, we first have to consider the nature of reinforcement.

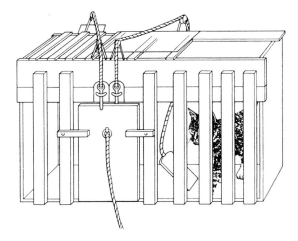

Figure 5.4. *A cat in one of Thorndike's puzzle boxes.*

There is a fundamental difference between the way a classical-conditioning experiment is conducted normally and the procedure used in an instrumental-learning experiment. In classical-conditioning experiments, a contingency is arranged between the CS (e.g. a bell) and the UCS (e.g. food), and the reinforcer (food) is applied regardless of the behaviour of the animal. In instrumental-learning experiments, the reinforcement (e.g. food) is contingent upon some particular behaviour of the animal (e.g. pressing a latch). Thus, in classical-conditioning experiments a contingency is arranged between a stimulus and an outcome, while in instrumental-learning experiments the contingency is arranged between a response and an outcome (Mackintosh 1974). However, these differences do not necessarily imply that different kinds of learning are involved in the two types of experiments. They do suggest, however, that different processes of reinforcement are involved.

In Pavlov's view, the occurrence of a reinforcing stimulus in a particular context will cause the responses elicited by that reinforcer to occur in anticipation of the delivery of the reinforcer. It is evident, however, that reinforcement is not always necessary for the occurrence of learned associations among stimuli. This can be seen most clearly by examining the phenomenon called 'sensory preconditioning'. The procedure is to present two conditional stimuli (CS_1 and CS_2) together for a number of trials before introducing the UCS. Thus, joint presentations of CS_1 and CS_2 are followed by pairing of CS_1 with the UCS. In the final test phase, the strength of the CR in response to CS_2 is measured.

The first clear demonstration of sensory preconditioning is that of Brogden (1939), who gave dogs 200 trials on which a light and a buzzer were presented simultaneously. One of these stimuli was then paired with electric shock to the paw to establish leg flexion as a CR. Test trials to the other CS produced an average of 9.4 CRs compared to only 0.5 CRs obtained from control experiments that did not involve prior pairing of buzzer and light. More recent experiments show that better results can be obtained if fewer pre-training trials are given and if the two CSs are presented not simultaneously but within a few seconds of each other (Mackintosh 1974).

The results of sensory-preconditioning experiments clearly show that pairing two neutral stimuli is sufficient to establish some learned association between them. It appears that if the stimuli are presented too often, some habituation occurs and learning does not improve. Quite apart from the fact that these results cannot be explained satisfactorily by a stimulus–response theory of learning, it is apparent that reinforcement is not necessary for associations to occur between two neutral stimuli. Pavlovian reinforcement is, therefore, not a necessary condition for the formation of associations, although it does facilitate their formation.

We now turn to the question of instrumental reinforcement. Thorndike's law of effect became the mainstay of the behaviourist's approach to animal learning. An extreme position was reached by Harvard behaviourist Burrhus Frederic Skinner who turned the law of effect into a definition of reinforcement. For Skinner, a reinforcer was any event that, if made contingent upon some aspect of behaviour, would cause the behaviour to increase in frequency. Skinner (1938) also assumed that any reinforcer could strengthen any response in the presence of any stimulus, provided that the stimulus could be sensed by the animal and that the response was within its capacity. Thus, the response and reinforcer were regarded as being essentially arbitrary, and this was a widespread view among learning theorists up to the end of the 1950s.

Skinner believed that all behaviour was modified and effectively controlled by the prevailing reinforcement contingencies. The idea that animal behaviour could be manipulated entirely by appropriate schedules of reinforcement represented the extreme behaviourist view (Skinner 1938). Skinner's behaviourist philosophy brought about a revolution in experimental technique that has persisted to this day.

In place of the trial-by-trial procedure characteristic of classical conditioning and experiments using puzzle boxes or mazes, Skinner devised the 'free-operant' procedure in which the animal is allowed to indulge freely in various activities, while the experimenter attempts to manipulate the consequences. Rats and pigeons were chosen most frequently for this type of experiment, although many other animals including humans were also used. Operant conditioning consists essentially of training an animal to perform a task to obtain a reward. A rat may be required to press a bar, or a pigeon to peck an illuminated disc, called a key. The typical method of training is called shaping.

Let us consider the training of a pigeon to peck a key to obtain a food reward. A hungry pigeon is placed in a small cage, equipped with a mechanism for delivering grain, and a key at head height, as shown in Figure 5.5. This type of apparatus nowadays is called a 'Skinner box'. Delivery of food is normally

Figure 5.5. *A pigeon pecking a key in a Skinner box.*

signalled by a small light that illuminates the grain. Pigeons soon learn to associate the switching on of the light with the delivery of food, and approach the food mechanism and eat the grain whenever the light comes on. The next stage of shaping is to make food delivery contingent upon some aspect of the animal's behaviour. It is usual to require the pigeon to peck the key, but Skinner claimed that any response could be shaped and that pigeons could be taught to preen or turn in small circles to obtain reward. Key pecking may be shaped by limiting rewards to movements that become progressively more similar to a peck at the key. Thus, when it has learned to approach a key for reward, the pigeon is rewarded only if it stands upright with its head near the key. At this stage, the pigeon usually pecks at the key of its own accord, but it can be encouraged by temporarily gluing a grain of wheat to the key. When the pigeon pecks the key, it closes a sensitive switch in an electronic circuit that causes food to be delivered automatically. From this point on, the pigeon is rewarded only when it pecks the key and the manual control of reward is no longer required. The animal is now ready for use in an experiment.

Skinner's approach depends upon the efficacy of reinforcement in modifying behaviour. His claim that any activity can be modified is illustrated by the various games that pigeons can be trained to play. Thus, Skinner (1958) reports that 'the pigeon was to send a wooden ball down a miniature alley towards a set of toy pins by swiping the ball with a sharp sideward movement of the beak. The result amazed us...The spectacle so impressed Keller Breland that he gave up a promising career in psychology and went into the commercial production of behaviour'.

Ironically, it was two of Skinner's students, Breland and Breland (1961) who first cast doubt upon the assumption that any activity could be modified by reinforcement. They found that, in attempting to train animals to perform various tricks, some activities appeared to be resistant to modification by reinforcement. For example, they attempted to train a pig to insert a coin into a piggy bank. The pig would pick up a wooden token, but instead of dropping it into the container, it would repeatedly drop it on the floor, 'root it, drop it again, root it along the way, pick it up, toss it in the air, drop it, root it some more, and so on' (Breland and Breland 1961). Similarly, they encountered chickens that insisted on scratching at the ground when they were supposed to stand on a platform for 10 – 12 seconds to receive a food reward. Subsequently, there have been numerous reports of this type. The Breland results violate Skinner's (1938) principle of least effort by which animals are supposed to obtain their rewards in the quickest and most comfortable manner. In all cases, reward is delayed considerably by the misbehaviour of the animals in their studies. Subsequent research has down-played the importance of stimulus–response learning, and has led to the recognition that much apparent instrumental learning is due to Pavlovian conditioning (see McFarland 1999, for a review). Moreover, there is evidence that associative learning involves some element of cognition (Dickinson 1980).

Cognition in animals

Cognition refers to the mental processes that cannot be observed directly in animals but for which there is, nevertheless, scientific evidence. Suppose we allow hungry pigeons to observe food presentations that are accompanied by the illumination of a small electric light. During the observation period, the pigeons are not allowed to approach the light or the food. Other (control) pigeons observe light and food presentations that are unrelated in time. At the end of the initial observation period, the pigeons are allowed to approach the light and food stimuli. All the pigeons tend to peck at the food delivery mechanism. However, the experimental pigeons also peck at the light, while the control pigeons do not. This results shows that the experimental pigeons must have formed a mental association between the food and the light, even though their behaviour during the initial phase of the experiment was the same as that of the control pigeons.

So far, we have seen that animals can learn to associate two events if the relationship between them conforms to what we normally call a causal relationship. Some of the conditions under which associative learning occurs are not consistent with the traditional view that animal learning is an automaton-like association of stimulus and response (S–R learning). There are two basic alternatives that have a long history, but are still considered important:
1. The view that animals can make associations between different stimuli (S–S learning).

2. The view that animals are designed to acquire knowledge about various relationships in their environment.

First of all, we have to distinguish two types of knowledge: 'knowledge how' and 'knowledge that'.

What do we normally mean by 'knowledge how'? We mean knowing as a matter of knowing how to do something, such as how to swim, how to ride a bicycle. This type of knowledge cannot be transferred from one task to another, and cannot be articulated. Thus knowing how to ride a bicycle involves a type of knowledge that cannot be used for anything except riding a bicycle, and that cannot be transferred to another person by speech or writing.

Riding a bicycle involves a procedure that is implicit, involving no explicit representations that can be transferred to another task or person. This type of knowledge is generally called 'procedural knowledge'.

What do we normally mean by 'knowledge that'? We mean knowing as a matter of having accessible, in different ways, usable information about an object, a person, a place, etc. In studies of humans, this has usually been called 'declarative knowledge', being knowledge that people declare that they possess, or declare to be the case. In this sense, declarative knowledge can only be had by language-using agents.

One can summarize this by saying that 'knowledge how' is (somehow) in, or part of, the system, in contrast to 'knowledge that', which is available for the system to work on (Pylyshyn 1984).

Animals do not have language abilities, and so cannot have declarative knowledge. However, it is perfectly possible that some animals may have the non-linguistic equivalent of declarative knowledge, and be able to deploy such knowledge in a variety of tasks. To allow for this possibility Lanz and McFarland (1995) revised the traditional terminology as follows:

- Procedural knowledge is knowing how. It is knowledge that is tied to a procedure. It cannot, therefore, be used in another procedure, or be accessed by another process. Dickinson (1985) equates procedural knowledge in rats with habits.
- Explicit knowledge involves explicit representations of facts that are (by their definition of explicit representation) accessible to many processes. Explicit knowledge, therefore, involves tokens, which represent 'facts'.

Because declarative knowledge (that people declare) is available only to language-using agents, it is a variety of explicit knowledge that pertains to humans, and possibly to artificial agents in the future. Where Dickinson (1985), and others who have experimented on animals, use the term 'declarative knowledge', Lanz and McFarland substitute the term 'explicit knowledge'.

The empirical issue in animal behaviour research is whether or not animals possess explicit representations that permit them to manipulate explicit knowledge. An animal that did not have such representations would only have procedural knowledge. It could learn (e.g. by forming S–S associations), but it could not think.

In a procedural system the form of representation — implicit representation (Lanz and McFarland 1995) — directly reflects the use to which the knowledge will be put. For example, Holland (1977) exposed rats to a tone–food relationship by occasionally presenting an 8 second tone and then delivering food pellets into a receptacle. Holland noticed that the rats developed a tendency to approach the food receptacle during the tone. This observation might suggest that during learning, the procedure of approaching the food receptacle is established, so that the learned information is stored in a form that is related closely to its use. The alternative possibility is that the rat learns that the tone causes the food, and thus establishes an explicit representation. The observation that the rat tends to approach the food receptacle during the tone then would have to be accounted for in some other way, because an explicit representation is passive in the sense that it does not control the animal's behaviour. The procedural representation thus provides a more parsimonious explanation of the rat's behaviour.

Suppose, however, that after the tone–food association is established, the rats are exposed to a food–illness relationship to the point where they refuse to eat the food when it is presented. The rats now will have formed two separate associations, tone–food and food–illness. The question is whether they are capable of integrating the two. On the one hand, a procedural account of learning would imply that the rats should not be able to integrate the two procedures, which have no factors in common. An explicit system, on the other hand, provides a basis for integration because both representations have the food term in common. Holland and Straub (1979) showed that rats can integrate information from such associations learned at different times. Rats exposed to a food-illness following tone–food showed a disinclination to approach the food receptacle when the tone was presented again.

The assumption that a procedural system cannot account for certain phenomena of animal learning is, of course, an assumption. It may be that an animal devoid of any cognitive ability could demonstrate such phenomena, but we are unable to test this possibility on animals, because we do not know of any animal that we can say, for certain, is devoid of the ability to manipulate explicit representations. A possibility is that robots can be used to investigate this question. Specifically the question is whether aspects of animal behaviour that are thought to be due to cognition can be performed by robots that are known (by the designer) to have no cognitive ability. This is the subject of our next case study.

Case study 8: Robotic experiments on rat instrumental learning *by Emmet Spier*

Animal learning

There are phenomena in animal learning theory that provoke controversy about the type and complexity of the cognitive machinery necessary for producing the observed behaviour. One such is the outcome devaluation effect, first described by Krieckhaus and Wolf (1968). Since then much work, mainly involving rat experiments, has been expended on understanding this phenomenon (for example, Holland and Straub 1979; Adams and Dickinson 1981; Colwill and Rescorla 1985). Typically these experiments involve training a rat in a Skinner box to press a lever for one particular reinforcer and to pull a chain for another, distinguishable, one. The reinforcers would generally be food items such as glucose solution or food pellets and the actions the rat makes to cause them to be delivered are called instrumental or operant actions. Once the rat reaches a satisfactory performance criterion it is then taken to a separate room, and, after being allowed free access to consume a selected reinforcer, it is then injected with lithium chloride solution (LiCl). This will cause the rat to become sick, and allows the exploitation of a phenomenon known as food aversion conditioning (Garcia *et al.* 1955), where rats avoid consuming a particular food after a close temporal experience with the food and a sickness. After recovery, the rats are then tested (without reinforcement) in the Skinner box to see if they will lever press or chain pull. Half of a four-group balanced experimental design (this one following Adams and Dickinson 1981) is illustrated in Figure CS 8.1.

Lever-press → Reinforcer 1 Chain-pull → Reinforcer 2	Reinforcer 1 → LiCl	Lever-press v Chain-pull?
Lever-press → Reinforcer 1 Chain-pull → Reinforcer 2	Reinforcer 2 → LiCl	Lever-press v Chain-pull?

Figure CS 8.1. *Outcome devaluation experiment.* This shows a balanced experimental design (following Adams and Dickinson 1981) for two groups of rats, with the other two groups being those that receive the opposite instrumental contingencies

Both Adams and Dickinson (1981) and Colwill and Rescorla (1985) found that the rat's response to the manipulandum associated with the delivery of the 'devalued' reinforcer was significantly attenuated. In the above account the 'outcome devaluation' was facilitated by food aversion conditioning; the phenomena have also been described under the same name in the context of a specific satiety, when the animal is simply fed to satiation with the relevant substance or resource (Colwill and Rescorla 1985; Shipley and Colwill 1996). The results are broadly similar.

These results are particularly interesting because at the very least they indicate that there must be some hidden state within the rats that can change between the two Skinner box sessions, one where they would activate the manipulandum and another, under exactly the same external physical conditions, where they would not.

Dickinson (1980 and subsequently 1989) gives a detailed consideration of possible internal mechanisms that could exhibit such a behaviour. He outlines the two forms which he calls either procedural or declarative; where a procedural system follows a behaviourist (Watson 1919; Skinner 1938; Hull 1943) conception of a behavioural account and a declarative system follows a Tolman (1932) type cognitive conception. Concisely put, a procedural account assumes that the structure of the representation of the knowledge directly reflects the use to which it will be put, much like associative forms of knowledge.

According to Dickinson, declarative knowledge is represented in a form that corresponds to propositions about relationships between events in the world. According to Lanz and McFarland (1995), declarative knowledge (that people declare) is available only to language-using agents, which does not include rats. However, there is no reason why rats should not have the non-language equivalent of declarative knowledge (i.e. be capable of thought without language), which they call explicit knowledge. In the most pared down incarnation this could take the form of logical statements such as, 'lever-press causes food-delivery', where lever-press and food-delivery are symbols explicitly representing the concepts lever press and food delivery. Such propositions require some additional machinery to process the symbols and translate the appropriate ones into actions. One important diagnostic criterion for explicit knowledge is that the information can also be misread, whereas the hardwired nature of procedural knowledge permits no such interpretation. The advantage of an explicit or declarative system is that through the use of its explicit symbols it can integrate its knowledge to arrive at an action that the simple procedural stimulus–response system might not be able to produce.

Utilizing this integrative power of an explicit/declarative model makes accounting for outcome devaluation very easy. The rat can be considered to have formed three statements concerning the apparatus: 'lever-press causes food-pellet', 'chain-pull causes glucose-solution' and, after food aversion conditioning, 'food-pellet causes sickness'. A simple integration, utilizing some kind of machinery that joins the two statements containing the common symbol food-pellet, resolves to produce a further statement, 'lever-press causes sickness'. A comparison of the two resultant operant statements, 'lever-press causes sickness', and, 'chain-pull causes glucose-solution', with some kind of, say, hedonistic selection method will then permit the statement, 'chain-pull causes glucose' to activate some action programs, or at least actions at the same level of description as a procedural account.

Outcome devaluation seems to be much harder to account for from a simple stimulus–response procedural approach. The training period of the experiment might install such new reflexes (conditional responses, building upon the ideas in associative theory e.g. Shanks 1995) as, 'see lever, press lever', and 'see chain, pull chain' while relying on previously existing unconditional responses such as, 'see food, eat food'. After food aversion conditioning a further conditional response would be established, 'see food-pellet, avoid food-pellet'. With such a configuration of responses it is understandably difficult to conceive how the rat might be able to act according to the observed phenomenon. There seems to be no way of integrating the aversion-conditioned response with the operant responses to suppress them; since no reinforcement is supplied in the test session, such an agent would 'see lever, press lever' without hindrance. Should the rat use this variant of a stimulus–response model then it would be unable to exhibit the outcome devaluation phenomenon.

In light of such thought, Dickinson (1985) states the strong position:

> My colleagues and I have argued that teleological control of instrumental behaviour [of which outcome devaluation was his example] cannot be explained, at least at the psychological level, in terms of internal associations which have just excitatory or inhibitory properties. Rather, we argue the knowledge about the action-goal relation must be encoded in a propositional-like form so that it can be operated on by a practical inference process to generate the instrumental performance.

Here he makes two claims, the first being that 'just excitatory or inhibitory' properties cannot account for outcome devaluation, and the second that outcome devaluation must be explained by a model similar in kind to the declarative model described above. This case study shows that by the incorporation of simple learning rules into an already established procedural model (Spier and McFarland 1998) phenomena very like outcome devaluation can be exhibited.

A procedural model

The model is based upon ethological theory, which stems from work initiated two decades ago by Sibly and McFarland (1976), McFarland and Houston (1981), and Tovish (1982). This established the mathematical basis of the current work, in particular the important concepts of resource availability and accessibility, leading to a methodology of linking mechanistic accounts of behaviour to functional accounts involving optimal changes in the animal's state. Spier and McFarland (1996) developed the ethological theory into a model from which they could construct a decision rule. They show that this rule does

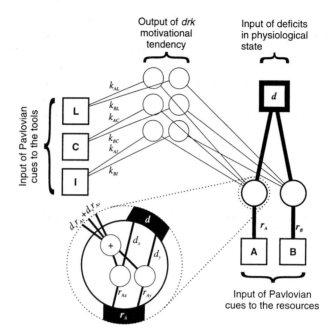

Figure CS 8.2. *The drk model depicted within the connectionist metaphor.* The thin-edged squares represent a scaleable cue (putatively Pavlovian) that acts as an input to both the r and k pathways. The thick-edged square represents the vector output of the physiological state space. The thin-edged circles represent a single computational unit which (unless otherwise indicated) multiplies its inputs. The slightly thicker-edged circles represent a computational group of which one example is drawn out in full, its output being the dr product. Thick lines represent a vector pathway and thin lines a scalar one.

better than various alternatives in a simple robot simulation. McFarland and Spier (1996) show that this rule makes functional sense in a real robot situation, and that the rule does better than alternatives, of comparable complexity, in these real robots, mainly by allowing the robot to be opportunistic. Spier and McFarland (1997) extend this rule in a particular direction (called tool using), and show that the amended rule still makes functional sense, and that it does better than the old rule in a simple robot simulation. This rule produces apparent (or implicit) planning (or appetitive behaviour), without any explicit representations. This rule, called the drk rule, forms the theoretical basis of the current discussion.

Figure CS 8.2 depicts the architecture of the model, which is more concisely expressed in matrix notation. Here we consider that an agent must satisfy two internal state variables, say h for hunger and t for thirst; this is expressed by the deficit vector d. In the case of an animat the 'physiological' state space would be those factors that the agent could measure that were correlated with its fundamental tasks. McFarland and Spier (1997) discuss a robot with a state space consisting of energy and work performed. From previous experience the agent has built a number of associations that link externally sensed resources to changes in its internal state (their availabilities), as shown in Figure CS 8.3.

As developed in Spier and McFarland (1996), the motivational tendency to perform any particular behaviour can be calculated by forming the matrix product [drk], where d represents the deficit state, r represents the resource availabilities and k represents the resource accessibilities (see Spier and McFarland [1996] for a full mathematical explanation). The r and k values are modulated by sensed (Pavlovian) cues to their presence, as elaborated later.

Resource availability r

reduction of deficit due to consumption of	food pellet	sucrose solution
hunger h	R_{Ph}	r_{Sh}
thirst t	r_{Pt}	r_{St}

Resource accessibility k

Tools at the animal's disposal	Lever-press	Chain-pull	Consummatory behaviour
For obtaining a food pellet	k_{PL}	K_{PC}	$k_{P\ agent}$
For obtaining some sucrose solution	k_{SL}	k_{SC}	$k_{S\ agent}$

Figure CS 8.3. *Resource availability and accessibility.* The resource availability elements represent the changes in the animat's state that result from consuming either a resource of type P (food pellet) or S (sugar solution). The resource accessibility elements represent associations between particular behaviours and how those behaviours influence the ease of obtaining any particular resource, where any particular k value can be considered to represent the long-term expectation of the particular reinforcer (P or S) that can be obtained by performing that particular behaviour (which we call a tool). Here L denotes a lever-press, C denotes a chain-pull, and 'agent' denotes the direct consummatory behaviour towards the resource. So we find that a single element in k, say k_{SL}, represents how the lever-press tool modifies the nominal accessibility of consuming the sugar solution; likewise, k_{PL} would represent the accessibility for the food pellet obtained by a lever-press.

A procedural account of outcome devaluation

Using the drk model as outlined above, we can now describe a possible configuration of the d, r, and k values that will be able to exhibit the outcome devaluation effect. With the model so described we can, in the situation where there are no free resources available, form calculations that can have their

194 Artificial ethology

magnitudes compared competitively (picking the largest) to assess which behaviour to perform. These are illustrated in Figure CS 8.4.

These values imply that lever pressing delivers food-pellets, that chain pulling delivers sugar-solution, and, from (a), that food-pellets only satisfy hunger whereas sugar-solution satisfies both thirst and, to a lesser extent, hunger.

If we consider a 'hungry' agent in which d = (20; 5), then [dr]P = 20 and [dr]S = 15; since $k_{PL} = k_{SC}$, there is no particular preference for either instrumental action, and so the agent would execute lever-press actions. Hence, incorporating the drk model into the 'head' of an agent will permit it to switch between lever pressing, chain pulling, and consummatory behaviour as might be appropriate.

In the case of food aversion learning we postulate that the animat can alter the appropriate vector in its r matrix via an association with the induced sickness. In defence of the abuses of a modeller in performing this act, all of the procedural and declarative accounts described above also assume such a sleight of hand step (and much worse) so it seems reasonable for the drk model also to make such a move. In our example we could imagine the situation portrayed in Figure CS 8.4c. Figure CS 8.4f shows how the animat's preferences are changed by the aversion conditioning. In this case the animat will chain-pull rather than lever-press.

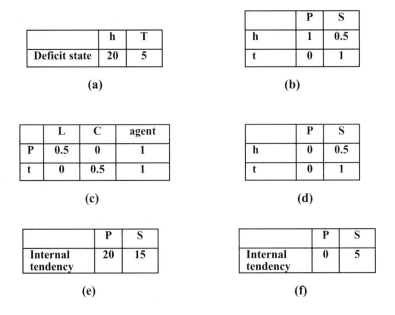

Figure CS 8.4. *Parameter values used in examples and experiments.* (a), (b), and (c) show the values of d, r, and k used in the example. (d) shows the values of r after lithium chloride treatment. (e) shows the matrix product dr from case (b), and (f) shows the matrix product dr from case (d).

In the case of the animal experiencing a specific satiety devaluation, the drk account for outcome devaluation is even simpler. In the above case, with d = (20; 5), if we feed the agent to satiation with food pellets then the agent's deficit vector would be d = (0; 5) so now [dr]P = 0 and [dr]S = 5; again, the agent will perform chain-pulling actions. This result is dependent on the resources having different values in the agent's state vector, and would not account for, say, differently coloured resources.

The behaviour described above is reminiscent of what would be expected in an outcome devaluation experiment, and the model is certainly described with the same (or even greater) rigour than those in Dickinson (1980). It is now necessary to consider where this model sits in the procedural-declarative classificatory scheme. The significant difference between the drk model and the straightforward procedural model outlined above is that both the appetitive/instrumental cue and the consummatory/reinforcing cue are not only associated but combined to form a conjoint cue strength in some common currency of the agent's motivation. From this combined cue the behaviour is determined; consequently, a change in either cue may control an instrumental act. Likewise, since the 'goal' of the behaviour has (through some postulated learning) shaped the r and k matrices but not been incorporated into them, it is not necessary for the goal to be 'encoded' in the agent for it to act in a goal-achieving manner. Inasmuch as a 'goal-achieving' system (McFarland and Bösser 1993) like the drk model can exhibit outcome devaluation, it would be incorrect to assume, as Dickinson (1985) has, that outcome devaluation is intrinsically a teleological phenomenon and necessarily requires a declarative type reasoning system for it to be made manifest.

Incorporating learning rules into the drk model

In the previous section it was assumed that the agent governed by the drk rules came pre-supplied with ready-made r and k matrices to provide it with the appropriate configurations to allow the agent to lever-press and chain-pull according to its motivational state. It might be reasonable to challenge the validity of such an assumption. Likewise, the supply of ready-made matrices could be seen as inserting some teleological control though the back door. For a fully procedural account, at the current level of description, it is also necessary to describe how the parameters in the r and k matrices can be generated.

In both the psychological and artificial intelligence literature, mechanisms for learning associations have often been based upon a simple error-correcting learning rule called the 'delta rule', first described by Widrow and Hoff (1960). Sutton and Barto (1981a) have shown that under certain assumptions the delta rule is formally equivalent to the Rescorla-Wagner (1972) learning theory developed to account for certain conditioning phenomena. This is attractive and important because it implies that there is a common set of basic assumptions underlying the learning rules in associative learning, and so it is not important to consider each minor variant as a distinct algorithm that needs its own special considerations. Indeed, it is well established that it is possible to incorporate learning into an animat with multiple tasks. Sutton and Barto (1981b) furnish an early example of animat-type reinforcement learning work while trying to demonstrate the psychological phenomenon of latent learning. The learning rules developed here were designed to be the simplest possible formulations that, while being compatible with some basic psychological observations, were also able to imitate only those learning requirements necessary for satisfactory performance of outcome devaluation.

The substantive decisions that were made in designing the learning rules are (1) to consider that the exponential learning asymptote of the availability r would be equal to the availability of the resource as experienced in *ad libitum* conditions. Likewise, (2) the accessibility k would capture the reduction of the resource's availability when it has to be obtained by way of a particular manipulandum, thus causing a reduction in the global availability. Thus we can, for r, use a simple delta learning rule whose learning rate is modified by the reinforcer's temporal distance (its 'eligibility' [Barto *et al.* 1983] captured in a time-decaying 'eligibility trace'). For k we can, additionally, take advantage of the decaying eligibility to capture, in k, the time constraint that manipulating the operant controls places upon the agent (since the action, say lever pressing, is temporally separated from obtaining the reinforcer).

Incorporating the learning rules into the drk model is a somewhat technical matter, fully described by Spier and McFarland (1998), and will not be elaborated here.

A reasonably full simulation of the outcome devaluation experiments requires some consideration of extinction. A common method used in learning experiments to ascertain the extent to which an animal has learnt a particular task is the use of an extinction trial. Here the animal would be placed in its Skinner box in the same conditions as those under which it was trained, but the effects of operating the manipulanda would be disabled. The animal would lever-press or chain-pull to no avail; such behaviours would diminish in frequency and eventually cease to occur. By comparing how much the animal would perform one of the unrewarded manipulative acts compared to the other act, the experimenter can determine which manipulative act was preferred by the animal. The quantity of manipulative acts performed also gives some measure of how well the association was learnt. Such an assessment could not be undertaken if the manipulanda were operative because it would be almost impossible to discern whether the animal's actions were being reinforced through the delivery of the food items or whether they were governed by what the animal had learnt.

It is not our aim here to provide an account of extinction learning. Neither is extinction learning necessary for our account of outcome devaluation. It is simply a diagnostic technique that experimenters use. It is also a complex phenomenon where some time later the animal may recommence the extinguished behaviour (known as spontaneous recovery). For our purposes and in the spirit of the simple account we have previously developed, extinction is modelled by the subtraction of a small quantity from the active k matrix element when any particular action is completed. For instance, when the agent lever presses for sugar solution then k_{SL} is reduced by a small quantity. This would have the effect of making the particular behaviour less likely to occur unless it was reinforced. Of course, with this permanent change in the k value there is no room for phenomena such as spontaneous recovery but this simple approach will permit a comparison of the data presented in the outcome devaluation studies with the simulations performed later.

Simulation

The previous sections have provided a description of the drk model that offers an opportunity to account for the outcome devaluation phenomena by taking a particular instantiation of the drk model and attaching some simple learning rules which have been developed to capture the kinds of values that such an account requires. We now look at a computer simulation experiment that set out to see if the previously discussed mathematical description of the drk model can actually generate its expected behaviour when placed in a less homogeneous environment than that of a mathematical model (Spier and McFarland 1997).

The simulation environment consisted of a continuous rectangular surface of 2000x2000 units within which resided the agent, the operant controls (manipulanda), a food hopper, and two types of reinforcer. The animat possessed a radius of 60 units and the resources one of 15 units. Figure CS 8.5 shows a graphical depiction of the actual simulation environment used. The various manipulanda are labelled LP (for lever-press), CP (for chain-pull), and IC (for irrelevant-cue), an extra manipulandum to provide some spurious cue signals in case the behaviour was brittle to the number of salient cues in the environment. In addition FH, the food hopper, which was also a salient cue, demarked an area of the arena into which the reinforcers could be delivered at some random position. The basic simulation design factors were as follows: time is measured in discrete cycles; the agent can move a finite but small distance each cycle; objects are recognized perfectly; the cue strength of any object is a linearly discounted function from a maximum to zero at the maximum sensor range; and any consummatory or manipulative action occupies one cycle.

The animat possessed eight different possible behaviours: do nothing, eat resource A, eat resource B, move five units in a specified direction, lever-press, chain-pull, operate the irrelevant tool, and operate the food hopper. The various manipulandum-directed actions only had an effect upon the simulated world if

198 Artificial ethology

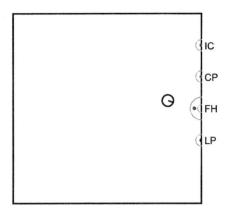

Figure CS 8.5. *An example of the 2000×2000 simulation environment.* The agent is depicted by the circle and radius; three manipulanda are shown (LP, CP, and IC) as well as the food hopper (FH) where, as can be seen, the reinforcers are deposited after an appropriate action. (For clarity the object sizes are not to scale.)

the agent was within ten units of the particular manipulandum. An implication of this partitioning of behaviours is that the agent had already ascertained how to use the various tools in the environment. Although a significant assumption, for the purposes of demonstrating the outcome devaluation effect in the context of the alternative models, this assumption seems reasonable since the alternative models also make it. In the experiments reported below, the operation of the irrelevant cue and the food hopper had no effect. The operation of the lever object caused one particular reinforcer to be delivered, say resource A, and the chain object caused the delivery of the other, say resource B. After an appropriate manipulandum was activated the resource would be delivered at some random location within the outer semicircle of the food hopper depicted in Figure CS 8.5. This random positioning would inject a small element of unpredictability into the decision-making algorithm of the animat since sometimes it would consume a resource closer to one tool than the other. Each of the two resources, A and B, supplied orthogonal changes in state with A satisfying only hunger and B satisfying only thirst (consequently we stop using the corresponding notation of P and S since, at least, sucrose solution satisfies both hunger and thirst). In this case, the change in deficit (h; t) from consuming resource A was (1; 0) and the change from B was (0; 1). Although the simple orthogonal case is described here, the simulation environment and the decision-making algorithm could have used non-orthogonal state changes from resources. However, the behaviour of the animat would be harder to interpret since it would exhibit progressively more 'trading-off' the greater the non-orthogonality of the resources.

The decision-making algorithm is in line with the initial description of the animat. It possesses a physiological state space of two dimensions, hunger (h) and thirst (t). Thus the two resources, food pellets (A) and glucose solution (B), supplied their benefits in terms of changes in hunger and thirst. So we have a situation similar to that depicted in Figure CS 8.4b.

For the tool matrix k we need to consider five tools: lever pressing (L), chain pulling (C), using the irrelevant cue (I), using the food hopper (F), and direct consumption by the animat (agent). This gives a situation similar to that depicted in Figure CS 8.4c.

There are ten possible appropriate products that the agent can form to evaluate a motivational choice, because the agent can attempt to consume either of the two resources using any of the five tools. Within the simulation environment, the r and k values are also adjusted by the cue strength (the extent to which the agent can perceive the various items). These, together with d, are multiplied to give the final calculation for the motivational strength of performing any particular behaviour. Cue values are calculated by taking the maximum value of either a linearly discounted r or k value with respect to the range the objects were from the agent (the model of cue strength), or a small ambient cue value that was set at the global expectation for encountering the particular object (the contextual cue).

For every cycle, each of these ten products would be calculated and the agent would act according to the product with the highest value. If the tool associated with this product was 'agent' then the animat would either move in the direction of the perceived resource, or consume the resource if it was adjacent to it. If the tool was a manipulandum then the agent would move in the perceived direction of the tool, unless it was adjacent to it when it would operate the manipulandum. Initially all the elements in the k matrix were set to a nominal value of 0.1. This small value was enough to initiate unfamiliar tool operations when the tendencies to perform known behaviours were low because of small deficits or absent cues to the appropriate tools or resources.

The procedure for testing such an animat was that it was first placed into the simulation environment and initiated. All the learning rules described above were active, adjusting the r and k matrix elements in the agent. After a period when it seemed that there was little change in the matrix values (the learning asymptote) then the outcome devaluation tests would be carried out. In the case of reinforcer devaluation (by LiCl in rats) then, as described above, the r matrix elements of the chosen reinforcer were reduced to zero after the agent had spent some time in an arena with no manipulanda present. This was done to prevent associations with the reinforcer devaluation and previous behaviour. As discussed previously, this somewhat magical act is equivalent to those made in all other associative or learning accounts. In the case of a specific-satiety devaluation, the animat, again in the barren arena, had its state variables reduced by multiples of the resource benefit of the chosen reinforcer. Subsequent to either devaluation act the agent was replaced in the arena with the manipulanda present, but deactivated, to assess the agent's performance in extinction tests.

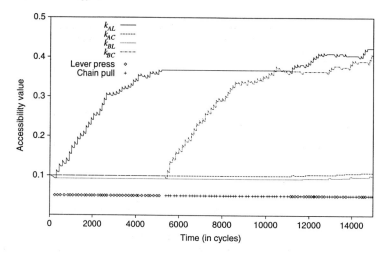

Figure CS 8.6. *An example of a typical accessibility (k) training session.* A lever-press (LP) delivers resource A and a chain-pull (CP) delivers resource B. Here, as in the text, k_{BL} would stand for the accessibility of obtaining a resource B using a lever-press. The bullets running parallel with the x-axis denote instants when a particular tool was operated (diamonds for A and crosses for B). The traces for the irrelevant cue (IC) are not shown here but are the same as k_{AC} and k_{BL}'s. Further discussion is supplied in the main text.

Results

The simulations were carried out in two phases. The first stage was to train the agent in the Skinner box so that it could learn the appropriate associations with its k values. The second stage was to test the agent in both the reinforcer devaluation and satiation situations.

Figure CS 8.6 shows a typical learning curve for the agent when a lever-press delivers resource A and a chain-pull delivers resource B. At the beginning of training, only one of the manipulanda was present at any time (the others being 'withdrawn'). The main reason for this was that the naive agent would have nominal accessibility values for all the tools. However, once one tool had a beneficial association it would be used in preference to all other tools, thus preventing the agent from sampling the rest of the environment. Of course, eventually the agent's deficits would be fully satiated in a particular dimension and the agent would find the motivational tendency to use that particular tool lower than its as yet unexplored alternative options. However, such a situation would only occur after an extended period of time. This procedure sped up learning all the key associations; the latter third of Figure CS 8.6 shows the agent finishing off its training session with all the manipulanda operative. As can be seen, the k values follow a negatively accelerated path to their asymptote and the irrelevant k values remain at their nominal values.

Motivation and learning 201

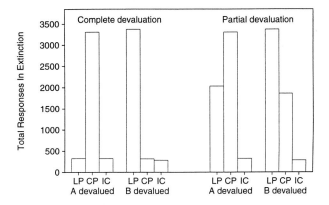

Figure CS 8.7. *Results of a simulated reinforcer devaluation extinction test.* Each of the sets of three bars represents a different experimental condition. The bar heights indicate how many attempts to operate each manipulandum (LP, CP, and IC) were executed. The leftmost two sets experienced a 'complete devaluation' where the r value of the chosen reinforcer was devalued to 0. The rightmost set experienced a 'partial devaluation' to 0.2.

Figure CS 8.7 shows the results of two different reinforcer devaluation experiments. The height of the bars denotes how many attempts were made to operate a particular manipulandum before no further attempts to execute actions against manipulanda were made. The 'complete devaluation' trials shown on the left are when the agent, with the training associations established in Figure CS 8.6, had one of its r values for its resources reduced to 0; Figure CS 8.7 shows the situations when either resource A or resource B is devalued. The 'partial devaluation' trials shown on the right are from a similar experimental set when the agent had one of its r values reduced to 0.2. As can be seen, in the 'partial

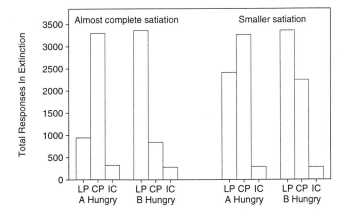

Figure CS 8.8. *Results of a simulated reinforcer satiation extinction test.* Each of the three sets of bars represents a different experimental condition. In the leftmost two sets the agent experienced an 'almost complete satiation' where the chosen deficit state was reduced to a deficit of 20. The rightmost set experienced a 'smaller satiation' to a deficit of 50.

devaluation' tests the number of attempts to operate the devalued manipulandum was notably greater than in the 'complete devaluation' case.

Figure CS 8.8 shows the results of a similar set of experiments when, instead of the reinforcer being devalued, the agent had one of its two state variables reduced significantly to simulate a satiation of that particular deficit. The pretrained agent from Figure CS 8.6 had its initial deficit levels set to d = (200, 200). The leftmost pair of graphs shows the results when the chosen deficit variable was reduced to a value of 20 and the rightmost pair shows results from when the chosen deficit variable was reduced to a value of 50. Again, the smaller the devaluation (satiation), the more the devalued action is executed.

Discussion

Figure CS 8.6 demonstrated that the extensions made to the drk model permitted it to learn appropriate k values (and the r values, although this has not been shown) from the agent's interactions with the environment. The results shown in Figures CS 8.7 and 8.8 clearly demonstrate that the drk model can exhibit outcome devaluation with both reinforcer devaluation and specific-satiation. Figure CS 8.9 shows data taken from a paper by Colwill and Rescorla (1985) demonstrating both types of outcome devaluation in rats. The data have been transformed to allow easy comparison with the simulation results presented

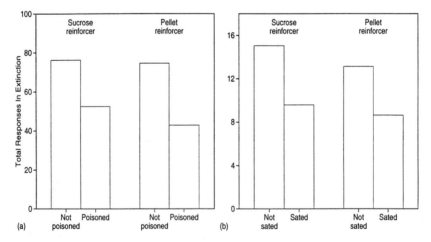

Figure CS 8.9. *Outcome devaluation in rats.* These graphs show data extracted and transformed from experiments 2 and 3 of Colwill and Rescorla (1985). (a) shows an outcome devaluation experiment using LiCl as a poison. The left hand side shows the number of responses to the tool that was rewarded by sucrose in training, and the right shows responses to the tool that was rewarded by Noyes pellets. (b) shows a reinforcer-specific satiety experiment where the animal was fed to satiety in one of the reinforcers. The left hand side shows the number of responses to the tool rewarded by sucrose in training and the right shows responses to the tool rewarded by Noyes pellets. Both graphs (a) and (b) show that the manipulandum associated with the devalued reinforcer is operated less than the alternative.

here; the shape of all the corresponding data sets, simulated and experimental, looks similar.

Of course, we cannot expect the simulation data to match exactly the rat data presented in Figure CS 8.9. There are too many unknown parameters, and rats are notably more complex than the drk model. However, the drk perspective may be able to explain the large difference in the number of responses Colwill and Rescorla's rats made when extinguishing under poisoning or satiation. Colwill and Rescorla attribute the reduced activity in the satiation experiment to a 'very global response suppressive effect' of the satiation process; however, from a motivational perspective we might be inclined to say that the two reinforcers were not particularly orthogonal to each other, so satiation with one reinforcer also has a satiation effect with another. This offers an immediately testable situation where a drk perspective would predict that there would be a less suppressive effect upon resources that could be independently shown to have a greater orthogonality in the animal's state space.

Less interesting reasons for differences between the two data sets could be attributed to the fact that the simulated agent only has eight possible behaviours and all are focused upon consuming the reinforcers; should it have several other behaviour systems in operation then the stark response seen in Figures CS 8.7 and 8.8 might have been more muted. Also, the model of extinction, as previously discussed, was very simple and did not interact with the mechanism of behavioural choice (except through the updating of associations); the agent therefore could not utilize its experience of being thwarted (or being surprised) to modify its actions in the short term.

The question now arises, whether or not the model presented above would actually satisfy someone who would hold a cognitive position such as in Dickinson (1985). The work of providing the explanation of outcome devaluation falls on two aspects of the drk model. The first is that each term (e.g. lever-press, food-pellet) is not just an object with which to form relations, but also possesses a value, innate or learned, to the animal expressed in terms of the change in the animal's state space. This aspect, although common in ethological ideas, is uncommon in psychology. (However, neural network models can implicitly use such information encoded within their weights.) Even so it seems a perfectly reasonable assumption in this context and additionally has the already established benefit of allowing both trade-off and opportunism in behaviour (Spier and McFarland 1996). The second is that associations with the reinforcer are incorporated twice into the model: firstly as the r component of the drk product; and secondly in the k matrix elements. The r association can be seen to be the classically conditioned aspect of the model with the k component being the operant conditioned component within a two-process theory that has a tradition within psychology stretching back to Hull. Neither would it be reasonable to argue that these additional assumptions provide the model with the 'more complex and powerful forms' of representations that

Dickinson (1980) accepts would probably be able to account for outcome devaluation in a procedural manner. Rather, coming from a functional viewpoint has provided a new perspective on the kind of information that these representations could carry and this provides an opportunity for a simpler explanation than postulating some cognitive capacities.

In summary, the most important thing to note is that this study has shown that a phenomenon for which there has been dispute over the requirement of cognition for its manifestation certainly does not necessarily require any cognition. Whether or not cognition is actually used by animals in the production of such behaviours is still an open question.

Comment on the use of robots

The procedural model presented in this case study manages to explain the outcome devaluation effect using the same set of initial assumptions as Dickinson's cognitive model (1985). However, certain assumptions may be more plausible within one framework than the other. For instance, both assume that the animals possess established (Pavlovian) categories for the various reinforcers and manipulanda. Although from a cognitive perspective this capacity is taken for granted (indeed, it is quite fundamental), many of the associative models do not wish to make such an assumption. One of the fundamental weaknesses of the associative model presented in this case study is that it uses a pre-wired architecture between established categories.

One way to address such a weakness is to extend the scope of the model by incorporating the capacity for it to form its own categories. Category formation requires raw sense data and such data (especially visual) are very difficult to model or simulate if, as is the case here, the animal can move within the environment. It seems as if the next step in substantiating a procedural model for the outcome devaluation effect requires a physical robot with visual input devices that instantiates a version of the drk model along with a suitably designed categorization scheme. This method forces one to address the questions of how to identify a category, and how to integrate such categories into the associative framework (for clearly, there is more than one way that they may be used). These are exactly the issues that remain to be addressed in the current drk model. In a real robot, the decisions of the drk model must be translated into a finely grained sequence of movements, yet retain the flexibility that is characteristic of drk decision making. This is another area where the issues connected with the deployment of real robots challenge the existing know-how.

End of Case Study 8

Representation and knowledge

We have seen that distinguishing among procedural, explicit, and declarative representations leads to hypotheses about animal cognition that can be tested on robots. Briefly, a declarative representation is a representation of knowledge about something. It is knowledge that such and such is the case, and can be declared to be the case. Because animals cannot use language to declare things, the animal equivalent of a declarative representation is an explicit representation. Knowledge how to perform some behaviour involves procedural representations, a set of instructions relating to some procedure. In a declarative or explicit system, knowledge is represented in a form that corresponds to a statement or proposition describing a relationship among states or events in the world. This form of representation does not commit the agent to use the information in any particular way. In a procedural system, however, the form of representation directly reflects the use to which the knowledge will be put.

If a human subject is told that A is bigger than B, and B is bigger than C, and then asked if A is bigger than C, we would expect a normal adult human to be able to infer that C is smaller than A, from the information provided. Such a problem is called a transitive inference problem. To us it seems logical that if A>B>C, then A>C, and a person capable of logical thinking should be able to attain the correct answer to this type of question. It is usually taken for granted that the ability to solve such problems involves manipulation of declarative representations.

Many transitivity experiments have been carried out on human adults and children, but McGonigle and Chalmers (1986) point to two main problems with the majority of such experiments. The first is that people learn an evaluative ordering more readily from better to worse than from worse to better. Consequently, when asking a question of the type 'if x is smaller than y, etc.', it is difficult not to introduce an incongruity between the form of the question and the subject's directional bias. The second problem concerns 'mental distance'. This concept implies that the more distant the items to be compared 'in the mind's eye', the faster decisions are made. Such results have been well established by Trabasso and Riley (1975), but it is not clear what these phenomena imply for the nature of mental representation.

McGonigle and Chalmers (1986) suggest that these phenomena may be due to pre-logical structures, in which case they may be present in subjects that are incapable of using conventional logical procedures. To circumvent these methodological problems, they suggest two types of experimental strategy. Firstly, to deal with the congruity issue, they suggest an alternative paradigm known as 'internal psychophysics' (Moyer 1973). This requires subjects to decide as rapidly as possible the serial relationships between objects from memory alone. For example, subjects might be presented with a pair of names of animals such as 'hen' or 'elephant' and asked to denote the larger by pressing

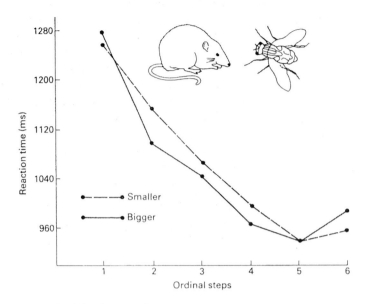

Figure 5.6. *Symbolic distance effect produced by 6-year-old children comparing sizes of animals on an ordinal scale. The pictures used, such as the mouse and the insect shown here, were drawn to be the same standard size.* (After McGonigle and Chalmers 1986.)

a switch below a panel bearing a printed name, or a picture (scaled to a standard size). 'As the knowledge representation is assumed to be established prior to the task, the problem of congruity endemic to the logical task can be eliminated. Now there is no basis for ambiguity. The degree of mapping achieved has to be between the question and the representation, and not between the question, the informing statement and the representation.' (McGonigle and Chalmers 1986).

Second, to investigate the question of pre-logical structures, McGonigle and Chalmers (1986) conducted experiments with younger subjects, 'whose failures in logical tasks have been documented (e.g. Inhelder and Piaget 1964) and monkeys (not well-known for their logical skills).' (McGonigle and Chalmers 1986, p146). They carried out experiments in which 6- and 9-year-old children were required to compare the sizes of familiar animals, presented either as written names (lexical mode), or as pictures (pictorial mode) of a standard size, as described above. They measured the time taken to compare symbols, and they also required the children to verify statements of size relationship in the conventional manner (e.g. is a cow smaller than a cat?). Their results with the conventional methods show the 'symbolic distance effect' obtained by Trabasso and Riley (1975). That is, the time taken to compare stimuli varies inversely with the distance between the stimuli along the dimension being judged. Along the size dimension, therefore, the time taken to compare the relative sizes of 'cat' and 'whale' is less than taken for 'cat' versus 'fox'. This is the type of result usually obtained with adults (Moyer 1973; Paivio 1975).

Similar results were obtained when 6-year-old children were presented with pictures scaled to equal size, as illustrated in Figure 5.6. Overall, McGonigle and Chalmers (1984) found that children as young as six show a significant symbolic distance effect in both pictorial and lexical modes when the simple comparative question (bigger or smaller?) is used in the test. They also found marked categorical asymmetry, particularly in the lexical mode. Not only was the time taken to judge an item as 'big' faster than that to judge one as 'small', but even for items judged as 'small' it was faster to deny that they were 'big' than to affirm that they were 'small'.

Figure 5.7. *Squirrel monkey (a) and child (b) performing similar transitivity tasks.*

McGonigle and Chalmers (1986) report a series of experiments on squirrel monkeys, designed to test their abilities in transitive inference problems, in a situation similar to that used for testing children (see Figure 5.7). In one experiment five monkeys were required to learn a series of conditional size discriminations such that within a series of size objects (ABCDE) they had to choose the larger or largest one of a pair or triad if, say, the objects were black; if white, they had to choose the smaller or smallest one (McGonigle and Chalmers 1980). They found that there was a significant and consistent effect of direction of processing such that decisions following the 'instruction' to find the bigger were made faster than those following the signal to find the smaller. Through practice, the animals became progressively faster, yet the absolute difference between the 'instruction' conditions remained invariant. Figure 5.8 summarizes some of the results of this experiment, and compares them with the results of similar experiments on children.

In another experiment, based on a modification of a five-term series problem given to very young children by Bryant and Trabasso (1971), they tested monkeys on transitive inference tasks in which the animals were trained to solve a series of four discrimination problems. Each monkey was confronted with a pair of differently coloured containers that varied in weight (A>B). When B was

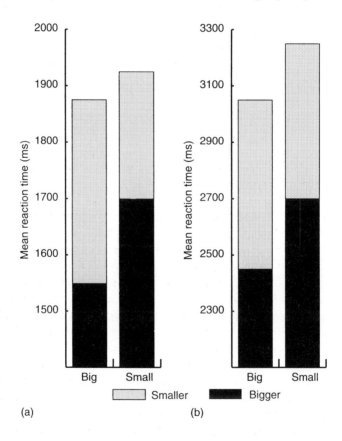

Figure 5.8. *Categorical and contrastive effects produced by (a) monkeys and (b) children.* (After McGonigle and Chalmers 1986.)

chosen reliably over A, the monkey moved to the next problem (B>C, where C must be chosen), and so on until the entire series was performed correctly. Only two weight values were used throughout the series, so no specific weight could be uniquely identified with the stimuli B, C or D. When the monkeys had learned to achieve a high level of performance on all four training pairs, regardless of presentation order, transitivity tests were given. In these, novel pairings were presented, representing all ten possibilities from the five-term series. The results showed impeccable transitivity, indistinguishable from the results obtained with 6-year-old children by Bryant and Trabasso (1971). Analysis of decision times revealed a significant distance effect in which the decision times for non-adjacent comparisons were significantly shorter than those for solving the training pairs, as shown in Figure 5.9.

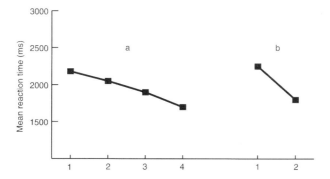

Figure 5.9. *Symbolic distance effect in monkeys.* (a) All comparisons, (b) not including comparisons between largest and smallest. The x-axis shows size differences on an ordinal scale. (After McGonigle and Chalmers 1986.)

On all major points of comparison, McGonigle and Chalmers (1986) found that the monkeys were identical in performance to young humans. Similar profiles in 6-year-old children, using both non-verbal and verbal forms of the same task, have also been reported (McGonigle and Chalmers 1984). So neither the nature of the task within a species, nor comparison of performance between species, seems to affect the conclusion that the symbolic distance effect (usually taken as evidence of cognitive processing in human adults and older children), and asymmetry in the direction of encoding (a characteristic feature of human transitive inference) occur in subjects unable to perform formal logical tasks. McGonigle and Chalmers (1986) come to the reasonable conclusion that the ability to order items transitively is a pre-logical phenomenon. While it is clear that monkeys do not solve transitive inference problems by manipulating declarative representations, it seems likely that the serial ordering involved in some of the more complex problems could be represented explicitly. In other words, some animals may be capable of non-verbal thinking (McGonigle 1987). Evidence that this may well be the case comes from experiments in which children and monkeys are given complex serial search problems (McGonigle and Chalmers 1998). Some of these problems enable the subjects to develop their own stimulus classifications and search strategies. The evidence suggests that the subjects develop cognitive strategies that economize on time, effort, and memory load. Such an ability implies a certain type of cognitive architecture in which there is a degree of both modularity and flexibility. Such ideas can be tested out on robots as we see in the next case study.

Case study 9: Robotic experiments on complexity and cognition *by Brendan McGonigle*

The problem

In taking a complex systems stance, it is necessary to 'bite the bullet' and study complex systems in their own right if we are to discover some of the key system principles that underwrite the flexible and powerful sorts of adaptation well instantiated in humans, and evident in the behaviour of some other primates (see above). Traditionally, two main approaches have held sway. One is behaviour-based, and targets simple systems or reactive subsystems of complex systems (exemplified by the preceding material in this book). The virtue of this approach is that simple systems that exhibit a tight coupling between input and output are more transparent than complex ones. However, there is a cost to this approach in terms of its generality of application to complex systems as a whole. Whilst few will doubt that complex systems share many design features with simpler ones, it is now becoming clear that many significant features of complex systems can be understood only by studying complex systems in their own right. Certainly it is the case that research into simpler systems has signally failed to demonstrate how such systems extend or scale up to more complex ones either in evolution, human development, or most recently robotics (see McGonigle 1990; Brooks 1991a); McFarland and Bösser 1993; McGonigle and Chalmers 1996, 1998.

It may be that complex biological systems are not merely quantitatively different from simple ones, but have a radical difference in organization. Given persistent failures of traditional learning approaches to scale up and make any phylogenetic sense, there remains the question of how cognitive functioning is achieved in a restricted class of biological agents and what, if any, are the radically different features of complex systems that are unique to them? A traditional answer to such questions, at least as far as human agents are concerned, has come from the second major stance — a representational, cognitive approach operating at the symbolic level. The virtue of this approach is that it recognizes as its starting point some of the specific complexities of human cognition and cultural adaptation. However, a persistent problem with this second approach lies both in the (in)adequacy of characterization of this 'end state' of human cognitive growth and (thus) the causal determiners of its growth and development.

In the face of the problems of classical cognition and the debates about the nature of 'representation', many have decided (perhaps wisely!) to eschew traditional problems concerning the sorts of mediation devices that humans may use, perceiving such processes as a kind of optional extra or some emergent subjective baggage which merely deludes us into a belief that we operate in the main quite rationally based explicit rules of rational control. Yet it is clear that many cognitive achievements are real and highly adaptive. Whatever the role of 'symbols in the head' they do work as externalized currency in a scientific society. Take, for example, the case of maps of physical space — another contentious area! Whatever we have in our heads, there are maps in the real world, which have been culturally evolved and which have, like other symbol systems, significant adaptive advantages. Take any (accurate) map of a large-scale environment for example. This is a representation! And the adaptive value of such a device is clear. Who would want to learn all the streets of a strange city by trial and error when a map has so much to offer in terms of possible routes, locations, relative positions of key landmarks, etc. so that with the right kind of interpretation we can find places as economically as possible. In short, the map has high utility, and whilst its advantages are not so clear in relatively small-scale situations, it becomes progressively more advantageous the larger the scale of the environment under review.

And this brings us to a core feature of our cognitive stance — which is not classical but embodied — that cognitive organization evolves ontogenetically in complex systems and solves, and is in turn challenged at various stages of growth and modification by, new problems which emerge from its previous successes. This dynamic process, we would argue, leads inexorably (but not teleologically) to the solution of large-scale problems in real space, in real time — an adaptive space of progressively more powerful solutions which reflect a complex agent's life history as it not only learns, but learns to learn.

For both biology and robotics, both troubled in the past with cognitive theories which do them not much good at all, a core idea is that the classical stances of realism on the one hand, and idealism on the other, are mistaken characterizations of the problem. Instead, an embodied stance offers a concept of a co-evolving agent and environment leading to a mutual specification designed to avoid the extremes of the classical representational stance on the one hand and the idealist one on the other.

Engineering the niche and the robot together

Tightly coupled behaviours in conventional behaviour-based robots must usually be well interpreted by the designer if they are to work at all. Whilst information from typical infra-red sensors severely under-specifies the objects

212 Artificial ethology

that lie in the robot's path, the tight coupling between world and action enables avoidance behaviours (at least) to occur robustly. However, that sort of behaviour is non-directed, and not strictly task achieving. To make such a system task achieving requires a lot more. One problem is that there is no task or goal 'attractor' in an avoidance behaviour. Another is that a reactive system based on infra-red sensors which invariantly stop movement when activated cannot distinguish between a potential collision situation and one in which object contact might benefit the agent — as in predation and consummatory behaviour. Another is that with movement based essentially on how the robot is perturbed in its motion through a cluttered space, there is usually no log or history of where is has been and certainly no locative information per se as to how it might return to a location it once historically occupied.

A hearing robot

With these problems in mind we designed a hearing robot (Figure CS 9.1) which hunted for a sound source in its environment (Donnett and McGonigle 1991). A beacon transmitted on one of 17 sound frequency bands, and the robot could pick up the different frequencies. The scale of the environment was such that in theory at least, the longest distance that the robot could be away from the sound still left it within range. In practice, the robot was in an office environment, and we found it would stray off behind objects such as filing cabinets, which baffled the sound coming from the sound source.

We decided to improve the basic controller by installing a capacity for failure diagnosis, so that 'no signal detected' could be interpreted as a system failure

Figure CS 9.1. *The hearing robot.*

(action: check ears), or a tuning problem (action: scan frequency bins), or a baffle problem (action: hunt for sound at random). With these procedures the robot succeeded in finding the source in an environment where the intensity of sound was by no means linearly related to the distance of the robot from the source.

Moving towards the sound source created some new problems. The detected signals indicated how close the robot was to the sound source, which was potentially useful as we wanted the behaviour at terminus to register something 'consummatory'. However, as the robot approached the speaker transmitting the sound, its infra-red sensors would initiate avoidance behaviour. Yet we wanted docking! So we had to introduce a state-based evaluator of the putative avoidance signal such that when the robot was closing in on the sound source it overrode its tightly coupled infra-red avoidance reaction and pushed the (protected) surface of the sound source.

That solved, we had a new problem of control. If the space was at all cluttered, the collision avoidance behaviours dominated the 'move to the sound source' behaviours to the point where the machine was bogged down for a lot of the time merely executing avoidance, producing a rapid oscillation between avoidance and attraction to sound. We needed to smooth this out. So we had the robot construct its first functional map of the environment layout, with the idea of smoothing behaviour by reducing the robot's speed in areas known to be cluttered. This was not a conventional map with two orthogonal axes. Instead, it was based, as in many such animal adaptations, on routes from a constant starting point. Thus when moving from this point, the robot had to log and time-code its interrupts en route to the sound source. The resulting performance gave the robot (although blind and unable to see the whole layout at once) the ability to predict regions of clutter and gear its speed accordingly, using a look-up table we provided.

Although this was a very primitive robot, we learned a lot from those early experiences in system control and signal interpretation. The problem of oscillations between competing behaviours, and the ensuing time wasting, was one lesson. Another was the value of selective interpretation of the same signal as a function of where the robot is in its state space, and/or where it is in the physical work context, where quite different behaviours may be required in response to the same stimulus conditions.

The Edinburgh R2 robot

This is a compact autonomous robot (Figure CS 9.2) weighing less than 2.3 kg, suitable for use in restricted spaces. It possesses a rich set of sensors, and a Cartesian manipulator, which allow it to alter as well as explore its environment. Essentially a small wheeled platform of about 20 cm in diameter, with differential steering, the R2 features a parallel jaw gripper that moves vertically

Figure CS 9.2. *The Edinburgh R2 robot.*

on the front, and is capable of lifting almost a kilogram. The infra-red sensors can detect reflective objects at a range of between 20 cm and 100 cm, the exact range of detection being a function of the reflectance of the obstacle. The robot is also fitted with light sensors, mounted around its periphery, which return a value between 0 and 255. The figure shows the bump sensors, which surround the bottom of the base and line the inside of the gripper. The Imputer onboard vision system is made by VLSI Vision Ltd, and contains a camera, a frame grabber, an 8051 microprocessor, and an RS232 interface. The niche for the robot is an office-type environment.

The aim was to implement a three-layered robot based on a logical hierarchy of function; from a basic reactive layer, to navigation and locative competence, to visual identification. The basic reactive layer, concerned with obstacle avoidance and similar functions, was of standard design, and can be thought of as a set of basic reflexes.

For navigation, we used two main sources of information. Firstly, as a general orientational frame of reference, we used the light gradients in the environment; secondly, we used a dead-reckoning scheme based on the signals sent to the wheel actuators. With low-level sensing constraints (at this stage of development), the kinds of locative representations we could achieve were strictly limited. Whilst dead reckoning is a useful competence over small ranges, error is cumulative with distance travelled; so for large-scale orientation it is not feasible in itself. Accordingly, we used a combination of differential light compass (see Nehmzow and McGonigle 1994) and dead reckoning. The light compass was used essentially to correct cumulative error arising from the dead-reckoning system.

Our first implementation of the light compass relied on assumptions about the light source that were incorrect for the environment in which the R2 usually operates. The main assumption was that the light source was essentially a point light source an infinite distance away. This works well when the light source is the sun, and the navigation worked very well when the R2 was run outside in a sunny environment. However, the usual environment for the R2 is an office with windows, giving all sorts of reflections and refractions, plus neon lights which are neither point light sources nor an infinite distance away. The result of this is that our first implementation of the light compass only allowed successful navigation on an excursion of about 6 m — not sufficiently accurate for the task at hand.

It is easy enough to ensure that there is only one significant light source in the habitat (by turning on and off lights, opening and shutting blinds, etc.), and shadows can be avoided if the obstacles the robot must avoid are not too high. The non-linearity of the light gradient must be dealt with, however, if the robot is not to be forced either to travel only very short distances or always to return by the same path as it went out (where the errors cancel each other out to some extent).

The obvious solution to the non-linear light gradient problem is to use polar co-ordinates. Instead of updating x and y co-ordinates, the robot updates its distance from the light source and its angle to the light relative to the angle at its starting position. This assumes that the gradient depends merely on the distance from the light source.

There are two main problems with this approach. Firstly, the robot needs to know its initial distance from the light source. The robot could work this out by setting out at right angles to the light source and using dead reckoning to determine how much the direction of the light changes relative to the distance travelled. Simple trigonometry could then be used to determine the distance from the light source. At the moment this is not implemented on the robot because it would require the robot to go through this self-calibration at the beginning of every run even though in the vast majority of cases the robot starts at the same location each time and in the same lighting conditions. Instead, the initial distance of the light source is compiled into the robot. The effective distance of the light source cannot necessarily be measured by conventional means such as a tape measure, because the light source might be the sun. Therefore to determine the effective distance we developed a human/robot interactive procedure by which the robot moves along the light gradient storing various measures and then uploads this data to the on-board computer. The human then examines a display of the data, explores the effect of changing the estimate of the distance to the light source, and selects the value giving the best results. This procedure works exceptionally well.

The second problem is that light sources tend not to be perfect point light sources, but short line light sources. This means that the gradient appears flatter in some regions of the habitat than in others. This problem can be largely overcome by adjusting the navigation algorithm (see McGonigle and St Johnston 1995).

The initial task-related navigational competence required from the robot was that it should be able to leave from a fixed base (a workbench), locate a target, and lift the target using its gripper. The R2 therefore begins each run from a fixed point against the bench. This not only allows us to compile into the robot the initial distance from the fixed light source, but also provides a means of eliminating cumulative error in the orientation of the robot each time it revisits the bench. Before heading off to locate and pick up the next target, the R2 pushes against the bench, thus aligning itself to its original orientation. The angle is then reset, before R2 launches itself off.

We divided the subtask of locating targets into two distinct phases, each varying in the level of behavioural granularity required. As with animals, the first problem was to get within the region of the target site by coarse-grained means; once there, the agent could begin a much more fine-grained and detailed search within a much smaller region of space. To enable the second phase of the subtask, Nehmzow and McGonigle (1994) installed high-reflectance strips on each target, which extended the sensor range limits of the infra-red sensors. In this way, the robot could be attracted to a target, and prepare a docking and pick-up procedure. Accordingly, the task illustrates neatly the importance of mapping between task stages and robot states to enable the robot to do the right thing at the right time.

To prepare specifically for a visual identification competence based on object location, we took the locative procedures a stage further. Now, instead of merely identifying and picking up specific objects, the robot is to discover targets first, then transport them to a bench location where they are to be arranged in a row. This is to establish a working set of objects in close proximity and makes the task of visual comparison less time consuming. Given this competence, where the only restriction on the size of the set to be interrogated is the size of the workbench, it becomes possible to use a type of active vision, as described in McGonigle and St Johnston (1995). Each target object is a freestanding pole which supports a target array for visual analysis; when managed as described, visual processing is triggered by the robot's grasp of each pole (Figure CS 9.3). Visual processing also terminates when the pole is released by the robot's manipulators.

Task grammar

The layers of competence we have engineered are not the control procedures. Instead, we have maintained the concept of ordered states the robot can be in, and designed task grammars which enable many different behaviours to be derived from the same three-layered 'stack'. Of course the syntax of these grammars is still specified by the human designer. But that is not the point at this stage. More to the point is that the decomposition of competencies at the hardware/wetware level is now detached from another dimension of behaviour-based decomposition, i.e. the chaining or stringing of a number of units of behaviour together to achieve useful adaptive tasks. This allows us to derive a variety of possible behaviours from the agent. And even the form of a simple chain — at least initially — allows for both a variety of tasks and takes us to the domain of multitasking.

With a developed functional architecture capable of supporting a variety of competences within the same agent, and task grammars that enable a variety of different tasks to be implemented by recruiting a core of behaviours from the same repertoire and reusing these in different combinations, we have rapidly reached the point where hand-crafted control has reached its limits. Initially, as in many robotics laboratories, our means of dealing with multiple behaviours was to develop the system incrementally through the addition of successive

Figure CS 9.3. *The Edinburgh R2 robot sorting poles.* The robot's task is to find the five poles, transport them to its workbench, and arrange them in a line in a particular order.

layers of control from those installed first. However, this approach soon confronts severe problems. It requires the designer to specify the priorities of control in advance; and this puts a further strain on the designer to anticipate what is required in advance of the robot's own behaviour. Secondly, as each new layer of control must be added to its antecedent hierarchically, the flow of control is necessarily unidirectional. Whereas a layer in a typical subsumption architecture can only execute or allow a layer strictly below it to execute, a more flexible solution would enable the system to call or inhibit any other behaviour module at any point (Humphrys 1996). How this might work in an actual robot is currently part of our agenda and I shall now give a flavour of the programme that we are now developing on a variety of related fronts — each scaffolded on earlier robotic implementations, but now moving away from hand-crafted, 'interpreted' competences and towards self-organized, self-regulated learning of the sort that we see as a core of primate growth and complexity in human and non-human alike (McGonigle and Chalmers 1996, 1998; McGonigle 1999).

The Nomad

The Nomad 200 (Figure CS 9.4) was purchased from Nomadic Technologies Inc. It has a diameter of 53 cm, is 96 cm high and weighs 59 kg with batteries. The drive system utilizes a three-servo, three-wheel synchronous drive mechanical system which provides a non-holonomic (with zero gyro-radius) motion of the base and an independent rotation of the upper structure (the turret).

The robot is equipped with radio LAN (Local Area Network) capable of sending and receiving signals while running freely in the environment. It has an electronic compass sensor. Distances and angles turned by the wheel actuators provide information that can be used to build a dead-reckoning (odometry) system. Error tends to accumulate, so this must constantly be re-calibrated using other sensory information.

The robot can detect when it bumps into low obstacles. This works for walls and skirting but not for all types of obstacle. It is obviously a detection method of last resort. There are also 16 sonar sensors, useful for long-range distance measurement. There are 16 infra-red sensors, for short-range (under 30 cm) distance measurement. Special highly reflective strips can be attached to the objects to be retrieved in order to exaggerate their appearance against the background (this makes them visible from about 60 cm). There is a highly sensitive Imputer camera, which enables visual processing and has been used successfully with the Nomad for sorting objects.

The Nomad is now supporting several developments from the work carried out in the simpler robots as described above. The first is to engineer a variety of tasks using the functional architecture we have developed, which allows us to

Motivation and learning 219

Figure CS 9.4. *The Nomad 200 robot used in the study.*

separate basic competences from their specific implementation under the control of particular task grammars.

Some of the Nomad robot primitives (basic behaviour routines) have been developed over years of previous work. The functional architecture distinguishes between the installed general layers of lower level competences within the system, and the task-achieving, task-level control devices which can recruit behaviours to support quite different patterns of activity, of varying granularity, from this general resource. Different task-achieving behaviours can be derived from the same design primitives as described above. These behaviours can be changed without altering the robot itself. A diverse number of different action selection architectures could now be built and tested on top of this framework.

Action selection is now a real requirement as we can now run several task scenarios at once. Furthermore, we have in these tasks gradable measures of success; for example, on the pole sorting task a number of poles may be missed when several tasks are run at once. In addition, as each task-achieving behaviour is a string of units or segments of behaviour, it can be interrupted at segment boundaries, and then continued afterwards. For example, a more urgent behaviour may interrupt an ongoing one by only temporarily halting it. So as a further development from where we have got thus far — multiple parallel tasks of diverse sorts – it is possible to undertake an analysis of the control processes that enable a repertoire of adaptive behaviours to operate together in the same niche to improve the robot's adaptive fitness. Although this work is only in its infancy, I shall mention it as an illustration of how the new problems derive and indeed escalate from earlier successes.

I suggest that multi-goal, real-time control architectures raise a new set of issues which go beyond simple additions of single-task skills, or simple forms of scheduling or optimization. These issues are long familiar to animal researchers, but it is clear that adaptive agent researchers will someday have to face them too.

In the real world, animals are not able to carry out their tasks uninterrupted — and this will hold too in any unpredictable environment into which adaptive agents are sent. Tasks must be interrupted in response to sudden unpredicted threats or opportunities, and resumption may be only enabled at unpredictable moments and locations. Tasks cannot necessarily be abandoned and then resumed in the same state later — the entire situation may have been lost in the meantime. Hence it may be necessary for tasks to resist interruption, to a degree dependent on the investment already made in them, and the difficulty of resumption. Some ongoing tasks may never terminate, but must still be interruptible. Others need to be done once a day, but at no specific time, and so need to be highly postponable, yet not indefinitely so.

Much of this, of course, violates the basic assumptions on which research into single isolated tasks is based. If this work is actually aimed at robots that perform unmolested in controlled factory environments, this is not a problem. But some of this work is aimed at developing autonomous agents, and here perhaps the field is in danger of making a basic strategic error: researchers are working on single complex tasks, under, it seems, the assumption that these can be somehow added together later into an agent with multiple complex tasks. But perhaps no such development path exists. Perhaps we need to think of the whole creature right from the start (see Dennett 1978), and work on a path from multiple simple tasks towards multiple complex tasks.

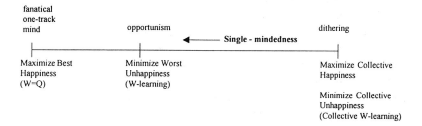

Figure CS 9.5. *A characterization of action selection schemes along the dimension of 'single mindedness'. (After Humphrys 1996.)*

Dealing with complexity

The flow of control necessary for multi-goal systems may be so complex that straightforward hand-design is impossible. Human programmers find it hard to think of the flow of control necessary in order to balance multiple long- and short-term goals, which may be of diverse importance and potential for opportunism, partially satisfied, potentially non-terminating, running over different time scales, and constantly interrupting each other and picking up control at opportunistic moments. Even if we manage to program smoothly integrated short-term interrupts, long-term patterns are very hard to ensure (for example, we can postpone low-priority goals, but it is easy to design a system in which, over the course of the day, we notice that low-priority goals are actually never attended to). Robot programmers are much less comfortable if they cannot have a single isolated problem, which can run unmolested to a successful terminating conclusion.

The solution, though, is not just to design restricted systems. It is to get some help with the system, some kind of automatic self-modification, given some broad rules that we can lay down. To do this, we have turned to an interesting class of models developed in theory and simulation by Mark Humphrys. Humphrys (1996) notes that action selection schemes, when translated into precise algorithms, typically involve considerable design effort and tuning of parameters. One possible approach to automating design is learning from success (rewards). Humphrys considers eight different approaches, ranging from centralized and co-operative to decentralized and selfish. They are tested in an artificial world, and their performance, memory requirements, and reactiveness are compared.

In designing action selection schemes, there is a tension between those criteria that contribute to several goals at once, and those that promote persistence. Humphrys (1996) formalizes this dilemma (see Figure CS 9.5) and devises metrics for comparing different action selection schemes. He comes to the conclusion that a characteristic identified as 'single-mindedness' pays off, because it promotes opportunism. In addition, schemes that rely on learning do better than 'innate' schemes.

Towards cognitive economy

The exact tasks chosen for a robot development program are not as important as having a good mix of different types of task, from slow-burning long-term ones which can be done at any time, to urgent but infrequent tasks; from tasks that depend on external stimuli to ones dependent on internal state. It is hoped therefore that the solution to the problem of managing one collection of tasks will address issues that will also occur in other quite different task collections. Moreover, learning based on an initially unknown task structure is likely to make the results more extendible to a wider range of collections of task. Thus we expect there to be basic principles associated with complexity of lifestyle *per se*.

One such principle is that of cognitive economy and search. Our many years of research with monkeys and children (see above), involving combinatorial and search problems, suggest similarities with the ordering problems inherent in managing action itself (Terrace and McGonigle 1994; McGonigle and Chalmers 1996). Few meaningful adaptive behaviours are single, tightly coupled ones. As in the case of words, it usually takes an ordered string of them to make any sort of adaptive sense. However, given this requirement, constraints of various sorts are necessary, because combinatorial explosion begins with relatively few elements. It is here that we see the main influence of organizational change both phylogenetically and ontogenetically. In the case of simple systems, such constraints can come well engineered, but at the price of plasticity. In the case of more complex, more plastic agents, however, there would appear to be a space of permissible variation — as in syntactic structures — which is well indicated by the flexible recombinative ordering seen in advanced primates, especially humans, yet which is well below the full permutative set of possibilities. In this context, whilst insects, for example, may lend themselves well to characterizations as finite state automata, complex systems do not.

The full dimensions of what seems to be a qualitative shift into new forms of dynamic organization in complex systems remains obscure: I do believe, however, that with the radical change in stance that we see now occurring, the way is open to new and exciting explorations of complexity based on a synthetic approach which places robotic and biological systems firmly within the same agenda.

End of Case Study 9

We have seen that there is an empirical issue in animal behaviour research, as to whether or not animals possess explicit representations and are able to manipulate such explicit knowledge. An animal that does not have such representations can only have procedural knowledge. It can learn, but it has no cognitive ability.

The ability to manipulate explicit representations is necessary for cognition. Indeed, some scientists go further, and insist that cognition requires symbol manipulation. Where the explicit knowledge is a proxy for an object, property or event, it is usually called a sign. There is a straightforward one-to-one relation between a sign and its referent (Hendriks-Jansen 1996). Symbols are not straightforward proxies for objects. When explicit knowledge takes the form of a symbol, it leads the agent to conceive of the object. What a symbol signifies is an act of conception, which is reactivated anew on each occurrence. As Pylyshyn (1984) points out, 'Organisms can respond selectively to properties of the environment that are not specifiable physically, such properties as being beautiful.' (p15).

Within the fields of classical artificial intelligence and robotics it is taken for granted that if you judge behaviour to be intelligent you are committed to see the behaviour as resulting from rational, mental, cognitive, symbolic processing. The view of classical artificial intelligence is that cognition involves two things: (1) There must be internal (mental) representations of the world (including aspects of the agents themselves), and (2) there have to be operations, or computations over these representations, to yield new beliefs, such as beliefs about necessary means to accomplish some goal. The representations are articulated in some sort of symbol system and, together, they make up the knowledge base from which an agent is supposed to reason and to decide what to do next.

This view is very different from the behaviour-based view, which insists on empirical evidence for mechanisms that are supposed to control behaviour. Within the field of AI, intelligent behaviour is seen in terms of sense-think-act cycles. The relevant internal processing is conceived in terms of symbol manipulation in which a program delineates a series of instructions that effectively specify the information processing to be carried out by the machine. The aim is to write for every kind of interesting intelligent behaviour a program that, when implemented on a machine, enables the machine to exhibit the relevant intelligent behaviour.

Behaviour-based robotics takes a very different view, and is much more biologically based. Firstly, the notion of intelligence is very different from that of AI (McFarland and Bösser 1993). Secondly, the philosophical outlook is very different (Hendriks-Jansen 1996). Thirdly, classical AI is widely considered not to have lived up to its promises, and the behaviour-based approach has been more successful in building robots (Brooks 1986, 1989, 1991b). However, this does not mean that we should abandon the possibility of cognitive processes. It may well be that some animals do have genuine cognition, and it may well be that cognitive process can be useful in robots. Indeed, the robot is an ideal tool with which to investigate these matters.

Chapter 6

Why robots?

In this chapter we address the more general questions relating to the use of robots in biological research. Much of the material in this chapter comes from discussions arising at the Artificial Ethology Workshop in 1998.

During the discussions we attempted to answer the following questions:

1. When is the use of real robots necessary, rather than merely desirable?

2. When is the use of real robots desirable?

3. What are the disadvantages of using robots rather than animals or computer simulations?

4. What can robots be used for within biology?

5. What details of the robot matter, and what do not matter?

6. What are the indirect benefits of using robots?

7. What are the dangers of using robots?

8. Are any specific scientific methodologies required in using robots?

9. What advances in robotics are needed to assist biological research?

Although there is clearly a degree of overlap between some of these questions, they provide a useful way of structuring a discussion of the value of robots to biology.

1. When is the use of real robots necessary, rather than merely desirable?

A hypothesis about animal behaviour may be investigated by four basic techniques: by using animals; by mathematics; by computer simulation; and by using robots. Clearly, using a robot is necessary only when none of the others can be used.

The use of an animal may be impossible for technical or ethical reasons. Technical impossibility can take a number of forms. Perhaps the strongest case is where the animal is extinct; Prescott and Ibbotson (1997) used a robot in an attempt to reconstruct the behavioural architectures of pre-Cambrian animals. In research on vertebrates and some invertebrates (e.g. octopuses), robots may be necessary for ethical reasons. For example, in Brendan McGonigle's case study, we can imagine that experiments involving manipulation of cognitive structure become desirable, at some point in the research. These could not be carried out on live monkeys for ethical reasons, but they could be carried out on an adequate robot model.

The use of mathematics always involves abstraction, and it is not always possible to form an abstraction that captures the level of detail required for any particular hypothesis. Even where this is possible, the resulting equations may be intractable.

The use of simulation is probably the single most troublesome area. As noted in Chapter 2, a simulation that adequately represents the relationships between sensing, action, and the environment can have many advantages over robots, especially in time and flexibility. However, great care must be exercised to ensure that artefacts do not arise from the details of the simulation; inadequate granularity, and synchronous rather than asynchronous computation are notorious sources of trouble. But there remain many aspects of the physical world, whether they involve the robot or the environment, that are insufficiently characterized for fine-grained simulation, and in cases where this detail matters, simulation may give misleading results. In Frank Grasso's case study, for example, the robot is used partly as a tool to investigate the environment. The underwater robot disturbs the pattern of flow in a way that could not be simulated, and this disturbance has a critical effect on what is sensed. Thus the robot makes it possible to test various theories of odour detection and plume tracing.

An unusual but necessary role for robots in biology is illustrated by Simon Giszter's case study: the prime subject for study may be an animal, but a robot may be the only way of providing some necessary experimental manipulation. In Simon Giszter's case, the manipulation is of aspects of the physical environment affecting the forces acting on the frog's limb under certain conditions of position and movement, and so the robot is not explicitly modelling biological processes. However, in studying the relationships between animals, whether of the same or different species, the use of an autonomous or teleoperated robot to interact with an animal is now becoming feasible; for example, Vaughan *et al.* (1998) constructed a robot 'sheepdog' which was able to round up ducks. The use of robots in this way is an extension of the use of manually operated models (see Chapter 5) which has a long history; Tinbergen (1951) used model sticklebacks to provoke reactions in real sticklebacks, because he could manipulate the behaviour of the model to some extent.

2. When is the use of real robots desirable?

Even when their use is not necessary, robots may be preferred to other approaches. There are three main reasons for such a preference: it may yield better results; it may be easier, cheaper, or quicker; and it may make the work more accessible to students, other scientists, and the public.

There is a long history to the notion that a robot may yield results that are in some sense better than those obtained using techniques of analysis or simulation. There are in fact two possible sources of benefit related to this factor: the robot is a real physical system, and it is embedded in the real world. The first ensures that the system is forced to deal with real sensors, real effectors, and real time. In a simulation, sensors and effectors would either be idealized, or would deviate from the ideal in some specified stochastic fashion, almost always by the addition of random independent Gaussian noise sources to each sensor and effector. When a real robot is used, it is extremely unlikely that the noise affecting sensors and effectors will be either Gaussian or independent; a particular problem is that correlated noise can produce systematic errors manifested as behaviour. It is certainly arguable whether it is preferable to use a (real) robot which reflects only one instance of all the possible correlated and uncorrelated variations in sensors and effectors, rather than a simulation which represents some mean of all possible uncorrelated variations, but this argument in support of real robots seems to be unassailable. Indeed, some investigators have deliberately used low-quality components and low-quality assembly to increase the deviation from the ideal of their robot systems (Kaelbling 1991).

Whatever the merits of the real robot argument, the real world argument is much stronger. The real world is the same everywhere and at all times, in that it always contains the full range of physical effects in the correct quantities. But every simulation, even so-called physics-based simulations, contains only a limited and idealized representation of these effects. The danger is that a simulation will omit some factor that is present in the real world and that materially affects the outcome of the experiment.

A robot experiment may be preferred to a computer simulation because, in some situations, it may be less expensive, involving less time and work. It is difficult to generalize about the circumstances under which the robotic approach would be cheaper, because the costs associated with a robot depend very much on circumstances. For example, if a simple wheeled robot carrying some simple sensors is all that is required, then there are now many inexpensive types readily available; on the other hand, if a special purpose robot had to be designed from scratch, this would cost very much more. In either case, most robots are very different from general purpose computers, and often require the learning of new programming skills, although some more recent designs are built round miniaturized standard computers to avoid this (Winfield and Holland 2000). Furthermore, since robots contain moving parts, they may need regular examination and maintenance by specialists, representing additional costs. Whatever the cost of the robot, the cost of the environment in which the robot is to be run may also be significant; a typical robot arena may occupy 40 square metres or more, and may require much the same facilities as any other experimental laboratory. Yet another source of expense is the means of recording data from robot experiments, which can be comparable to the cost of gathering data from animal studies.

On the other hand, for a computer simulation, it is likely that low-cost access to adequate and well-maintained computing facilities will be available; there will be no additional requirement for space or data recording, and no need for the programmer to learn about a new system; the simulation can be run for any length of time, and at any time of the day or week; and it is likely, though not inevitable, that the simulated animal or robot will run much faster than the real equivalent. However, the computer must be programmed not only with the details of the simulated robot's control system, but also with the details of the environment and of the sensors' response to it, and of the effectors' effects on it. It is this programming, especially that concerned with the environment, that represents the potential extra cost of simulation. Of course, if the environment is very simple, and the sensing and actuation very straightforward, or if the programming has already been done for some other purpose, then the programming cost can be low. The instances when this cost becomes very high often centre around complex dynamic and poorly understood environments — for example, turbulent water containing chaotic plumes of odourant. Therefore, where the physical situation to be simulated is very complicated, as in the case studies of Grasso and Webb, it may be less expensive to use a real robot, than to attempt to simulate the physical world in the detail required.

3. What are the disadvantages of using robots rather than animals or computer simulations?

A major difficulty in using robots in biological research is the interdisciplinary nature of the enterprise. Work with robots requires expertise that is not readily available in the biological community. Collaboration with roboticists can be expensive and slow, at least in the early stages, as each discipline gradually learns about the requirements and constraints of the other. Of course, in time the magnitude of this disadvantage may diminish, as robots become more readily available and accessible, and as education in biological robotics improves. The *avant-garde* nature of research in this area may put off some scientists, but may be a challenge to others. A related problem is that biological research carried out using robots may be unfamiliar and difficult to evaluate for biologists with no knowledge of robotics, and may therefore encounter difficulties in finding acceptance in mainstream journals. Although there are now a number of journals at the interface between biology and robotics, they have as yet made little headway into the biological community.

Conducting experiments with real robots is constrained to occur in real time, whereas simulations can often be run faster than real time. For example, in Emmet Spier's case study, conducting the experiments with real robots would have taken many weeks, as was the case with the original animal experiments. The careful use of simulated agents allowed much more thorough investigation in the available time. It often seems plausible to assert that experimenting with robots uses about the same time and resources as experimenting with animals, but this is rarely the case, because additional time is almost always required to tune the robots' physical and computational parameters to provide an adequate model of the animals' behaviour. A further difficulty is that robots, like any other electromechanical systems, change with use and time; this change can be so rapid that it can require recalibration or even replacement of components within an extended set of experiments such as those described in Owen Holland's case study. Although animals may sicken or die within an experimental time frame, they do not usually wear out their components.

Animal experimentation, especially within psychology, has traditionally depended on replication as a check on the reality of experimental results. With equivalent animals, a given experimental manipulation is expected to produce consistent results on different occasions, allowing for the individual variability of the animals. (Equivalence might require the use of a particular laboratory strain, rather than using just any example of the species.) With robots, it may be necessary to use the same type of robots loaded with very similar programs to achieve this. But while it may be possible to acquire one or more robots equivalent to the robots used in some reported experiment, it is almost never possible to acquire sufficient information to replicate the program and parameter settings. (In fact, it is frequently impossible to replicate simulations for this reason; even where pseudocode is given, different implementations can often give different results.) Where it is impossible to obtain the same type of robot used in a given study, it is inevitable that there will be differences in the programming, and so it is difficult to interpret a failure to replicate the original results. However, successful replication under those conditions may indicate a real and strong effect; for example, as described in Owen Holland's case study, the puck clustering reported by Beckers *et al.* (1994) was replicated by Holland and Melhuish (1999) using different robots and a (necessarily) different program, but obtaining essentially the same result, giving confidence that it is a strong and robust phenomenon.

4. What can robots be used for within biology?

The most basic use for robots in biological research is as substitutes for animals. They can be used to test hypotheses about the behaviour of models of animals, as in Frank Grasso's case study. More commonly, they can be used in demonstrating the adequacy of a model or hypothesis (Holland, Webb, Spier, Cruse), and as a pointer to predictions that may be tested in real animals (Halperin, Giszter). They may also be heuristically valuable, since the requirements of building and working with robots can raise questions that can assist in structuring the approach to a behavioural problem domain (Ayers, McGonigle). Of course, all of these comments apply to any method of simulation; however, the use of a robot leads to a different kind of simulation (hardware simulation) and may lead to new areas of research and thinking. This is shown in Barbara Webb's study, where the use of a robot makes the problem transparent. In particular, by using a robot to represent an animal in the physical world, the experimenter can bring the same intuitions to bear as would be used when observing an animal.

However, robots have a variety of other uses. One of the most valuable is in exploring and characterizing the environment from the animal's vantage point, as in Frank Grasso's study. More broadly, robots may be used as sensor platforms to gather relevant information from locations that cannot easily be accessed using any other technology, such as fixed instrumentation; this is one of the intended end uses of the robot under development in the project described by Joseph Ayers.

Interaction with animals is also a useful role for a robot. This may take two forms. In the study mentioned in (1) above, Vaughan *et al.* (1998) used a robot to induce flocking behaviour in ducklings; the robot constituted an experimental manipulation of the ducklings' environment, and in some related experiments the flocking behaviour, rather than the robot, was the subject of study (Henderson 1999). This is similar in essence to Simon Giszter's case study, where the robot is used to manipulate the frog's environment, and the subject of study is the frog's behaviour. In contrast, Lund *et al.* (1997) used a robot cricket to locate a real singing cricket; the subject of study in this case was the robot itself.

A particularly informative use of robots is their use in studying problems concerning autonomy. Animals are autonomous, both in terms of energy and motivation (see McFarland 1999 p460), and robots can be very useful tools in investigating the constraints placed on a system by the requirement of autonomy. It is worth noting that the majority of robots in commercial use have severely limited autonomy, or achieve apparent autonomy within a highly engineered environment and a very circumscribed task situation; the study of autonomy usually requires the construction of new robots, requiring significant engineering involvement.

Robots also offer advantages that are only indirectly related to research. It is a fact of experience that many people find mobile robots fascinating. The roots of this phenomenon have not been studied, but one characteristic is that people tend to perceive the robots as intentional beings, and attribute all sorts of internal processes to them that are entirely inappropriate. (It should be said in passing that the attribution of similar characteristics to many animals is also inappropriate, but may have a similar origin within human psychology.) The use of robots can therefore attract a great deal of publicity, and hence funding, for research projects, regardless of the scientific merit of the work.

There is also some potential for using robots within biology as teaching aids, but this is undeveloped at present.

An interesting perspective is supplied by Janet Halperin's implementation of her biologically inspired system on TRex, the commercial excavator. Although the motivation was the solution of a technical problem in the real world, and perhaps the demonstration of the potential power of biologically inspired algorithms for commercial applications, the exercise does have a biological relevance, in that it shows that the biological strategy is flexible, and capable of generalization beyond the rather narrow niche – fighting fish behaviour – in which it was discovered.

5. What matters about robots?

When a robot is used as a substitute for an animal, or as a vehicle for investigating a model of animal behaviour, it is important both that it should capture the behaviour under investigation, and that it should not introduce new behaviours that might interfere with it or otherwise affect the results. There are as yet no explicit criteria for deciding whether these requirements have been satisfied; however, as long as these conditions are met, the scientific value of using a robot is unaffected by other considerations, such as shape. In Barbara Webb's study, for example, an anatomical feature (see Figure CS 2.4) of crickets was simulated electronically, and installed into a robot. The robot then played a role in behavioural experiments, in which it is irrelevant that the robot is anatomically unlike a cricket. In other cases, however, advantages may be gained from a certain amount of biomimicry. As Joseph Ayers points out, real lobsters have good hydrodynamic stability, and provide a proven design that might well be relevant in underwater robotics. Similarly, the design of water-current receptors can benefit from the lobster example.

Robots are of course made from different hardware to animals, and where the nature of the hardware matters, as for example in physiological research, robots are unlikely to find application. However, as noted in Chapter 2, a recent trend is the construction of robots using animal components; for example, it has proved possible to study how the silkworm moth navigates in relation to minute traces of silkworm moth pheromone by mounting isolated antennae on a robot, and using the electrical activity picked up from the antennae to control the robot's movements (Kuwana *et al.* 1996).

It is also worth noting that a robot does not operate in isolation – its behaviour is the outcome of its interaction with the environment. Just as robots are abstractions from real animals, the environments in which most robots operate are abstractions from the environments in which animals operate. For example, Barbara Webb's cricket robots are constrained by mechanical factors to operate in a typical indoor environment with a flat floor, rather than in a grassy field. In the context of the study, this is clearly not a crucial difference. However, in studies such as Frank Grasso's, the robot's environment is identical with the experimental environment in which the animal is studied; given the nature of his study, with its emphasis on environmental properties, this is a necessary constraint.

6. Indirect benefits of robots

The use of mathematics and simulation to model aspects of an animal's behaviour inevitably involves omitting many features of the situation; it is not always easy to decide which features are unimportant and need not be modelled. With a robot, on the other hand, it is not possible to omit quite so many features, because both the robot and the animal are constrained by embodiment in the real world. One is forced to consider a whole agent, not merely an abstraction of certain components of an animal's behaviour, and one is also forced to deal with all the problems surrounding the interface between the robot and the physical environment. This can entail a great deal of extra work, but it makes it very difficult to avoid issues that may be critical — especially when the behaviours under investigation have a strong sensory-motor involvement. A further benefit – and cost – of robot systems is that all relevant characteristics must be precisely specified at a fine level of detail, to a greater degree than is usually required when using a mathematical or simulation model.

Using a robot is also an excellent way of appreciating the problems facing the animal in the situation under study. For example, McGonigle puts the agent (animal, child, or robot) in a situation where it is faced with the task. With the robot, the process of attempting to bring it to perform the task, and the nature of the failures encountered, clearly reveal the problems involved, and indicate possible solutions. He found that if he introduced a hierarchical organization of rules, this was a useful way of cutting down the search space by limiting the number of admissible combinations (McGonigle and Chalmers 1998).

7. Dangers of robots

Apart from the obvious operational risks and dangers of using any electromechanical systems, there is a particular danger associated with robots that appear to mimic animal behaviour: a robot demonstration often produces a stronger impression on the observer than may be strictly merited. The key factor underlying this is the tendency mentioned in (5) for robots to be experienced as intentional entities, rather than simply being mobile pieces of hardware; this very strong effect, to which even biologists seem to be subject, tends to bias the observer towards seeing the robot as being closer to the animal than it is. There is also the risk that the positive impact of a demonstration can blunt the recognition that demonstration is potentially a weak form of proof; it can be as important to show that the robot fails to demonstrate the behaviour under certain circumstances.

8. Are any specific scientific methodologies required when using robots?

One will not go far wrong in using the same basic experimental methods whether using animals or robots, because robots share many of the characteristics of animals. Individual robots of a given type will differ slightly from one another, just like animals. Other things being equal, a given animal or robot will behave slightly differently on different trials; neither animals nor robots are statistically stationary, though not always from the same causes. Both animals and robots may react to 'irrelevant' cues, or to unsensed environmental factors that nevertheless affect sensing or movement. Multiple trials, replication, and careful experimental design avoiding systematic errors are therefore mandatory with both types of experimental subjects. In fact, the biologist is much better prepared for working with robots than is the computer scientist, because he or she has a knowledge of the techniques for dealing with intrinsic variability. The computer scientist might be tempted to see such variability as a temporary state of affairs, which will cease to exist when a new, improved robot design is developed.

Robots provide opportunities for using techniques that are difficult in animals. For example, the simultaneous logging of internal processes and external behaviour is trivial in most robot experiments, but can give unrivalled insight into the ways in which apparently complex observed behaviour is determined by relatively simple sensory inputs and internal processes, as in the studies by Emmet Spier and Frank Grasso.

9. What advances in robotics are needed to assist biological research?

Conventionally, robotics technologies can be divided into sensing, actuation, computation, and power issues. At first sight, the factors limiting biologists in using robots more extensively seem to centre around sensing and actuation; available computational resources far outstrip the abilities of biologists to model animal nervous systems at any level, and the current typical robot endurances of between 20 minutes and 2 hours seem adequate for most purposes.

Sensing is really an unsatisfactory term; it includes both the transducer function of the sensory apparatus, and the results of the processing of information from the transducer function. In some modalities, such as olfaction in moths, the transducer function is dominant: the antennae carry sensory cells which respond only to one specific type of pheromone molecule, and when they produce output, it means that the pheromone is present. In other modalities, such as vision, the transducer function is much less important than the processing function. It is a fair generalization to say that, with the emergence of nanotechnology, the transducer-based aspects of sensing are rapidly ceasing to be a constraint in robotics. With respect to the processing-based aspects, the situation is less clear. Where the nature of the processing carried out by an animal is well enough understood, it is often possible to mimic that processing using digital or analogue techniques. For example, Franceschini has built models of the fly's visual system both in software and in hardware (Franceschini *et al.* 1992); more recently, Harrison has produced an integrated circuit version of a similar system capable of being mounted on a small robot (Harrison and Koch 1998). However, when the basis of processing remains obscure, or when it is technically impossible to implement it in a form suitable for mounting on a robot, sensing is still a constraint.

Actuation refers to any means by which the robot exerts force on its own body or on the environment. In fact, the effects of actuation fall into two very distinct categories: moving through the world, and acting on it. The majority of robots studied within a biological context are primarily concerned with moving through the world, usually along a flat floor. In those that also act on the world, their action tends to be derived from moving through the world: for example, in Owen Holland's study, the robots push objects around as they move; in Brendan McGonigle's, the robot can pick an object up, move through the world, and put the object down. Although these activities seem rather limited, they are not constrained by current robot technology. As far as movement through the world is concerned, robots have been built that can walk on legs, swim like fish, wriggle like snakes, fly, and hover. Acting on the world is the sole purpose of most industrial robots, which can manipulate, weld, paint, glue, and so on. The apparently narrow range of activities in robots used within biology is rather a reflection of the fact that, for many questions of biological interest, it simply does not matter exactly how the robot gets about, or how it acts on its environment, and so the easiest and cheapest solution can be used. In cases where these are exactly what does matter, as in the work of Holk Cruse, Joseph Ayers, and Simon Giszter, existing robotic technologies seem to be adequate for the moment.

It is also possible to divide actuation up in a similar way to sensing. At the lowest level, there is a form of transduction — from some control input to some movement or force generation. Man-made structures tend to use a single transducer of an appropriate size and power to control each degree of freedom; biological structures tend to use multiple transducers in parallel, which introduces a different and more complex type of control problem. Man-made structures also tend to constrain the available degrees of freedom to simplify the problems of simultaneously controlling multiple actuators; biological structures typically have many more degrees of freedom than are strictly necessary, with a correspondingly more difficult control problem. There is thus a processing element in actuation, in addition to the transduction element. It is perhaps significant that two of the papers in this volume (those of Holk Cruse and Simon Giszter) are devoted to the investigation of how the nervous system approaches this processing; it is knowledge of this that is a constraint, rather than the unavailability of a suitable synthetic muscle.

The papers in this volume clearly show that robots have a place within biology, and specifically within ethology, and that they can contribute to the solution of problems in many different areas. Since what is under study is the behaviour of robots, we feel that it is quite reasonable to characterize the area as 'artificial ethology'. At present, this is clearly no more than an adjunct to natural ethology, but it is worth considering whether this will always be the case. Animals are currently the only fully autonomous behaviour-producing systems available for study. They all developed in the same way – through evolution acting on organic chemistry – and they all do the same thing — reproduce successfully. When we study animal behaviour, we study systems for producing behaviour, but these systems must be constrained in certain ways by the context in which they arose. Robots have the potential for allowing us to explore autonomous behaviour-producing systems that are constrained neither by evolution, nor by materials, nor by reproduction. If we had a complete knowledge of such systems – in other words, if a complete artificial ethology existed – then we might be able to appreciate how natural ethology, as a specific case, relates to the general case.

Bibliography

Adams, C.D. and Dickinson, A. (1981) Instrumental responding following reinforcer devaluation. *Q. J. Exp. Psychol.* **33B** 109–121.

Adler, H. E. (1971) Orientation: sensory basis. *Ann. N.Y. Acad. Sci.* **188** 1–408.

Allison, J. (1979) Demand economics and experimental psychology. *Behav. Sci.* **24** 403–415

Altermark, B., Isa, T., Lundberg, A., Pettersson, L.G., and Tantisira, B. (1993) Characteristics of target reaching in cats. II. Reaching to targets at different locations. *Exp. Brain Res.* **94** 287–294.

Arbib, M. and Cobas, A. (1991) Schemas for prey catching in frog and toad. In *From Animals to Animats: Proc. First Intl. Conf. on Simulation of Adaptive Behavior* (ed. J-A. Meyer and S.W. Wilson) pp. 142–151. MIT Press, Cambridge, MA.

Ayers, J. (1992) Desktop motion video for scientific image analysis. *Adv. Imaging* **7** 52–55.

Ayers, J. and Crisman, J. (1992) The lobster as a model for an omnidirectional robotic ambulation control architecture. In *Biological Neural Networks in Invertebrate Neuroethology and Robots* (ed. R. Beer, R. Ritzmann, and T. McKenna) pp. 287–316.

Ayers, J., Crisman, J., and Massa, D. (1992) A biologically-based controller for a shallow water walking machine *IEEE Proc. on Oceanic Systems* (1992) pp. 837–842.

Ayers, J. and Davis, W.J. (1977a) Neuronal control of locomotion in the lobster, *Homarus americanus*. I. Motor programs for forward and backward walking. *J. Comp. Physiol.* **115** 1–27.

Ayers, J. and Davis, W.J. (1977b) Neuronal control of locomotion in the lobster, *Homarus americanus*. II. Types of walking leg reflexes. *J. Comp. Physiol* **115** 29–46.

Ayers, J. and Davis, W.J. (1978) Neuronal control of locomotion in the lobster, *Homarus americanus*. III. Dynamic organization of walking leg reflexes. *J. Comp. Physiol.* **123** 289–298.

Ayers, J. and Fletcher, G. (1990) Color segmentation and motion analysis of biological image data on the Macintosh II. *Adv. Imaging* **5** 39–42.

Ayers, J., Zavracky, P., McGruer, N., Massa, D., Vorus, V., Mukherjee, R., and Currie, S. (1998) A modular behavioral-based architecture for biomimetic autonomous underwater robots. In *Proc. of the Autonomous Vehicles in Mine Countermeasures Symposium*. Naval Postgraduate School.

Baerends, G.P. (1941) Fortpflanzungsverhalten und Orientierung der Grabwespe *Ammophila campestris* Jur. *Tijdschr. Entomol.* **84** 68–275

Barto, A., Sutton, R., and Anderson, C. (1983). Neuronlike adaptive elements that can solve difficult learning control problems. *IEEE Trans. on Systems, Man, and Cybernetics*, **SMC-13**, 834–846.

Basil, J.A. and Atema, J. (1994) Lobster orientation in turbulent odor plumes: simultaneous measurement of tracing behavior and temporal odor patterns. *Biol. Bull.* **187** 272–273.

Bässler, U. (1976) Reversal of a reflex to a single motorneurone in the stick insect, *Carausius morosus*. *Biol. Cybern.* **24** 47–49.

Bässler, U. (1993) The femur-tibia control system of stick insects - a model system for the study of the neural basis of joint control. *Brain Res. Reviews* **18** 207–226.

Bässler, U., Rohrbacher, J., Karg, G., and Breutel, G. (1991) Interruption of searching movements of partly restrained front legs of stick insects, a modal situation for the start of a stance phase? *Biol. Cybern.* **65** 507–514.

Bässler, U. and Wegner, U. (1983) Motor output of the denervated thoracic ventral nerve cord in the stick insect, *Carausius morosus*. *J. Exp. Biol.* **105** 127–145.

Beckers, R., Holland, O.E., and Deneubourg, J.L. (1994) From local actions to global tasks: stigmergy and collective robotics. In *Artificial Life IV* (ed. R.A. Brooks and P. Maes) pp. 181–189. MIT Press, Cambridge, MA.

Beer, R.D. (1991) *Intelligence as Adaptive Behavior: An Experiment in Computational Neuroethology*. Academic Press, New York.

Beer, R.D., Chiel, H.J., and Sterling, L.S. (1993) Computer-simulated insects that adapt to their environment may be the next stage in the evolution of artifical intelligence. In *An Artificial Insect*. American Scientist, Cleveland, OH.

Berkinblit, M.B., Feldman, A.G., and Fookson, O.I. (1986) Adaptability of innate motor patterns and motor control mechanisms. *Behav. Brain. Sci.* **9** 585–638

Berkinblit, M.B., Feldman, A.G., and Fookson, O.I. (1989) Wiping reflex in the frog: Movement patterns, receptive fields, and blends. In *Visuomotor Coordination: Amphibians, Comparisons, Models, and Robots* (ed. J-P. Ewert and M.A. Arbib) pp. 615–630. Plenum, New York.

Bizzi, E., Giszter, S.F., Loeb, E., Mussa-Ivaldi, F.A., and Saltiel, P. (1995) Modular organization of motor behavior in the frog's spinal cord. *Trends Neurosci.* **18** 442–446.

Bizzi, E., Mussa-Ivaldi, F.A., and Giszter, S.F. (1991) Computations underlying the execution of movement: a novel biological perspective. *Science* **253** 287–291.

Bonabeau, E., Dorigo, M., and Theraulaz, G. (1999) Swarm *Intelligence: from Natural to Artificial Systems*. Oxford University Press, Oxford.

Bonabeau, E., Theraulaz, G., Deneubourg, J-L., and Camazine, S. (1997) Self-organization in social insects. *Trends Ecol. Evol.* **12:5** 188–193.

Bowerman, R.F. and Larimer, J.L. (1974a) Command fibres in the circumoesophageal connectives of crayfish I. Tonic fibres. *J. Exp. Biol.* **60** 95–117.

Bowerman, R.F. and Larimer, J.L. (1974b) Command fibres in the circumoesophageal connectives of crayfish II. Phasic fibres. *J. Exp. Biol.* **60** 119–134.

Braitenberg, V. (1984) *Vehicles: Experiments in Synthetic Psychology.* MIT Press, Cambridge, MA.

Breland, K. and Breland, M. (1961) The misbehaviour of organisms. *Am. Psychol.* **16** 661–664.

Brogden, W.J. (1939) Sensory pre-conditioning. *J. Exp. Psychol.* 25 323–332.

Brooks, R. (1986) A robust layered control system for a mobile robot. *IEEE J. Robotics and Automation*, **RA-2** April 14–23.

Brooks, R. (1989) A robot that walks; emergent behaviour from a carefully evolved network. *Neur. Computation* **1** 253–262.

Brooks, R.A. (1990) Elephants don't play chess. *Robotics and Auton. Syst.* **6** 3–15.

Brooks, R.A. (1991a) Challenges for complete creature architectures. In *From Animals to Animats* (ed. J-A. Meyer and S.W. Wilson) MIT Press, Cambridge, MA.

Brooks, R.A. (1991b) Intelligence without Representation. *Artif. Intelligence* **47** 139–160.

Brooks, R.A. (1991c) Intelligence without reason. *Int. Joint Conf. Artif. Intell.*, Sydney, Australia, 569–595.

Brooks, R.A. (1991d) New approaches to robotics. *Science* **253** 1227–1232.

Brooks, R.A. (1997). Intelligence without Representation (revised and extended version). In *Mind Design II: Philosophy, Psychology, Artificial Intelligence.* (ed. J.Haugeland) pp. 395–420. MIT Press, Cambridge, MA.

Brooks, R.A. (2000) From robot dreams to reality. *Nature* **406** 945–947.

Brunn, D. (1998) Cooperative mechanisms between leg joints of *Carausius morosus*. I. Non-spiking interneurones that contribute to interjoint co-ordination. *J. Neurophysiol.* **79** 2964–2976.

Brunn, D. Dean, J. (1994) Intersegmental and local interneurones in the metathorax of the stick insect, *Carausius morosus. J. Neurophysiol.* **72** 1208–1219.

Bryant, P.E. and Trabasso, T. (1971) Transitive inferences and memory in young children. *Nature* **232** 456–458.

Bullock, T. H. (1978) *An Introduction to Neuroscience.* Freeman, San Francisco.

Cattaert, D., El Manira, E., Marchand, A., and Clarac, F. (1990) Central control of the sensory afferent terminal from a leg chordotonal organ in crayfish *in vitro* preparation. *Neurosci. Lett.* **108** 81–87.

Chasserat, C. and Clarac, F. (1980). Interlimb coordinating factors during driven walking in crustacea. *J. Comp. Physiol.* **139** 293–306.

Chepelyugina, M.F. (1947) Ph.D. Thesis, University of Moscow.

Chrachri, A. and Clarac, F. (1989) Synaptic connections between motor neurones and interneurones in the fourth thoracic ganglion of the crayfish, Procambarus clarkii. J. Neurophysiol. **62(6)** 1237–1250.

Chrétien, L. (1996) Organisation spatiale du matériel provenant de l'excavation du nid chez Messor barbarus et des cadavres d'ouvrières chez Lasius niger (Hymenopterae: Formicidae). Ph.D. Thesis, Département de Biologie Animale, Université Libre de Bruxelles.

Clynes, M. (1961) Unidirectional rate sensitivity. A biocybernetic law of reflex and humoral systems as physiological channels of control and communication. Ann. N.Y. Acad. Sci., **93** 946–969.

Cohen, M.J. (1955) The function of receptors in the statocyst of the lobster (Homarus americanus). J. Physiol. **130** 9–33.

Colwill, R.M. and Rescorla, R.A. (1985). Postconditioning devaluation of a reinforcer affects instrumental responding. J. Exp. Psychol.: Anim. Behav. Proc. **11** 120–132.

Connell, J.H. (1990) Minimalist Mobile Robotics: a Colony Architecture for an Artificial Creature. Academic Press, San Diego CA.

Connell, J.H. (1992) SSS: a hybrid architecture applied to robot navigation. In Proc. 1992 IEEE Conf. on Robotics and Automation pp. 2719–2724.

Consi, T.R., Atema, J., Gouldey, C., Cho, J., and Chryssostomidis, C. (1994) AUV guidance with chemical signals. Proc. IEEE Symp. on Autonomous Underwater Vehicle Technology. Cambridge, MA July 19–20,1994.

Consi, T.R., Grasso, F., Mountain, D.C., and Atema, J. (1995) Explorations of turbulent odor plumes with an autonomous underwater robot. Biol. Bull. **189** 231–232.

Crisman, J.D. and Ayers, J. (1992) Implementation studies of a biologically-based controller for shallow water walking machine. SPIE Vol. 1831 Mobile Robots **VII**.

Cruse, H. (1976) The control of the body position of the stick insect (Carausius morosus), when walking over uneven surfaces. Biol. Cybern. **24** 25–33.

Cruse, H. (1990) What mechanisms coordinate leg movement in walking arthropods? Trends Neurosci. **13** 15–21.

Cruse, H. and Bartling, C. (1995) Movement of joint angles in the legs of a walking insect, Carausius morosus. J. Insect Physiol. **41** 761–771.

Cruse, H., Bartling, C., Dean, J., Kindermann, T., Schmitz, J., Schumm, M., and Wagner, H. (1996) Co-ordination in a six-legged walking system: simple solutions to complex problems by exploitation of physical properties. In From Animals to Animats 4 (ed. P. Maes, M.J. Mataric, J-A. Meyer, J. Pollock, and S.W. Wilson) pp. 84–93. MIT Press, Cambridge, MA.

Cruse, H., Kindermann, Th., Schumm, M., Dean, J., and Schmitz, J. (1998) Walknet - a biologically inspired network to control six-legged walking. In Neural Networks 11 (ed. R.A. Brooks and S. Grossberg) pp. 1435–1447.

Darwin, C. (1859) On the Origin of Species by Natural Selection or The Preservation of Favoured Races in the Struggle for Life. John Murray, London.

Davis, W.J. (1979) Behavioral hierarchies. *Trends Neurosci.* 5–8.

Davis, W.J. and Ayers, J. (1972) Locomotion: Control by positive feedback optokinetic responses. *Science* **177** 183–185.

Davis, W.J. and Kennedy, D. (1972) Command interneurones controlling swimmeret movements in the lobster. 1. Types of effects on motorneurones. *J. Neurophysiol.* **35** 1–12.

Davis, W.J., Mpitsos, G.J., and Pinneo, J.M. (1974) The behavior hierarchy of the mollusk Pleurobranchaea II. Hormonal suppression of feeding associated with egg-laying. *J. comp. physiol.* **90** 225–243.

Dawkins, R. (1996) *Climbing Mount Improbable*. Penguin Books, London.

Dawkins, R. (1974) Hierarchical organization: a candidate principle for ethology. In *Growing Points for Ethology* (ed. G. Bateson and R. Hinde) pp. 7–54. Cambridge University Press, Cambridge.

Dean, J. (1991a) A model of leg co-ordination in the stick insect *Carausius morosus*. I. A geometrical consideration of contralateral and ipsilateral co-ordination mechanisms between two adjacent legs. *Biol. Cybern.* **64** 393–402.

Dean, J. (1991b) A model of leg co-ordination in the stick insect *Carausius morosus*. II. Description of the kinematic model and simulation of normal step patterns. *Biol. Cybern.* **64** 403–411.

Dean, J. and Cruse, H. (1995) Motor pattern generation. In *Handbook of Brain Theory and Neural Networks* (ed. M.A. Arbib) pp. 600–605. MIT Press, Cambridge, MA.

Dean, J. and Wendler, G. (1982) Stick insects walking on a wheel: Perturbations induced by obstruction of leg protraction. *J. Comp. Physiol.*, **148** 195–207.

Delcomyn, F. (1980) Neural basis of rhythmic behavior in animals. *Science* **210** 492–498.

Delius, J.D. (1969) A stochastic analysis of the maintenence behaviour of the skylark. *Behaviour*, **33** 137–138.

Deneubourg, J.L., Goss, S., Franks, N., Sendova-Franks, A., Detrain, C., and Chrétien, L. (1991) The dynamics of collective sorting: Robot-like ants and ant-like robots. In *Simulation of Adaptive Behaviour: From Animals to Animats* (ed. J-A. Meyer and S.W. Wilson) pp. 356–363. MIT Press, Cambridge, MA.

Dennett, D.C. (1978) *Brainstorms*. Bradford Books, Vermont.

Dennett, D.C. (1995) *Darwin's Dangerous Idea: Evolution and the Meanings of Life*. Penguin Books, London.

Dickinson, A. (1980) *Contemporary Animal Learning Theory*. Cambridge University Press, Cambridge.

Dickinson, A. (1985) Actions and habits: the development of behavioural autonomy. *Phil. Trans. R. Soc. Lond. B* **308** 67–78.

Dickinson, A. (1989) Expectancy theory in animal conditioning. In *Contemporary Learning Theories: Pavlovian Conditioning and the Status of Traditional Learning Theory* (ed. S. Klein and R. Mower). LEA, NY.

Dickinson, A. and Dawson, G. (1989). Incentive learning and the motivational control of instrumental performance. *Q. J. Exp. Psychol.* **41B** 99–112.

Dittmer, K., Grasso, F.W., and Atema, J. (1995) Effects of varying plume turbulence on temporal concentration signals available to orienting lobsters. *Biol. Bull.* **189** 232–233.

Dittmer, K., Grasso, F.W., and Atema, J. (1996) Obstacles to flow produce distinctive patterns of odor dispersal on a scale that could be detected by marine animals. *Biol. Bull.* **191** 313–314.

Doherty, J.A. (1985) Trade-off phenomena in calling song recognition and phonotaxis in the cricket, *Gryllus bimaculatus* (Orthoptera Gryllidae). *J. Comp. Physiol. A.* **156** 787–801.

Donnett, J. and McGonigle, B. (1991) Evolving speed control in mobile robots. *Proc. Vision Interface Conf.*, Calgary.

Dreyfus, H.L. (1992) *What Computers still can't do: a Critique of Artificial Reason.* MIT Press, Cambridge, MA.

Eaton, R.C. and DiDomenico, R. (1985) Command and the neural causation of behavior: a theoretical analysis of the necessity and sufficiency paradigm. *Brain Behav. Evol.* 132–164.

Eilts-Grimm, K. and Weise, K. (1984) An electrical analogue model for frequency dependent lateral inhibition referring to the omega neurones in the auditory pathway of the cricket. *Biol. Cybern.* **51** 45–52.

Eustace, D., Barnes, D.P., and Gray, J.O. (1994) A behaviour synthesis architecture for cooperant mobile robot control. *Control-94: IEE Conference Publication 389.* IEE, London.

Evoy, W. and Ayers, (1982) Locomotion and control of limb movements. In *The Biology of Crustacea, Vol. 4. Neural Integration and Behavior* (ed. D.C. Sandeman and H. Atwood) pp. 62–106. Academic Press, New York.

Feigenbaum E. (1968) Artificial intelligence: themes in the second decade. *IFIP Congress '68*, Final Supplement p. J-11.

Ferrée, T.C., Marcotte B.A., and Lockery S.R. (1997). Neural network models chemotaxis in the nematode *Caenorhabditis elegans*. In *Advances in Neural Information Processing Systems 9* (ed. M.C. Mozer, Jordan M. I., and Petsche T.) pp. 55–61. MIT Press, Cambridge, MA.

Fleming, K.M., Reger, B.D., Sanguineti, V., Alford, S., and Mussa-Ivaldi, F.A. (2000) Connecting brains to robots: an artificial animal for the study of learning in vertebrate nervous systems. In *From Animals to Animats 6: Proc. 6th Int. Conf. on Simulation of Adaptive Behavior* (ed. J-A. Meyer, A. Berthoz, D. Floreano, H. Roitblat, and S.W. Wilson) pp. 61–72. MIT Press, Cambridge, MA.

Floreano, D. and Mondada, F. (1994) Automatic creation of an autonomous agent: Genetic evolution of a neural-network driven robot. In *From Animals to Animats 3: Proc. Third Int. Conf. on Simulation of Adaptive Behavior* (ed. D. Cliff, P. Husbands, J-A. Meyer, and S.W. Wilson) pp. 421–430. MIT Press, Cambridge, MA.

Floreano, D. and Mondada, F. (1996) Evolution of plastic neurocontrollers for situated agents. In *From Animals to Animats 4: Proc. SAB96* (ed. P. Maes, M.J. Mataric, J-A. Meyer, J. Pollock, and S.W. Wilson) pp. 402–410. MIT Press, Cambridge, MA.

Flynn, A. and Brooks, R.A. (1989) Battling reality. MIT AI Lab Memo 1148.

Fookson, O.I., Berkinblit, M.B., and Feldman, A.G. (1980) The spinal frog takes into account the scheme of its body during the wiping reflex. *Science* **209** 1261–1263.

Fraenkel, G.S. and Gunn, D.L. (1940) *The Orientation of Animals*. Clarendon Press, Oxford.

Franceschini, N., Pichon, J.M., and Blanes, C. (1992) From insect motion to robot vision. *Phil. Trans. Roy. Soc. B.* 337 283–294.

Franks, N.R. (1989) Army ants: a collective intelligence. *Am. Scient.* **77** 139–145.

Franks, N.R. and Sendova-Franks, A.B. (1992) Brood sorting by ants: distributing the workload over the work-surface. *Behav. Ecol. Sociobiol.* **30** 109–123.

Franks, N.R., Wilby, A., Silverman, B.W., and Tofts C. (1992) Self-organising nest construction in ants: sophisticated building by blind bulldozing. *Anim. Behav.* **44** 357–375.

French, M.J. (1988) *Invention and Evolution*. Cambridge University Press, Cambridge.

Frik, M., Guddat, M., Losch, D.C., and Karatas, M. (1998) Terrain adaptive control of the walking machine Tarry II. Proc. European Mechanics Colloquium Euromech 375: Biology and Technology of Walking, pp. 108–115.

Gandolfo, F. and Mussa-Ivaldi, M.A. (1993) Vector summation of end-point impedance in kinematically redundant manipulators. Proc. IEEE/RSJ Int. Conf. Intelligent Robots and Systems (IROS93). **3** 1627–1634.

Garcia, J., Kimmeldorf, D.J., and Koelling, R.A. (1955) Conditioned aversion to saccharin resulting from exposure to gamma radiation. *Science* 122 157–158.

Giszter, S.F. (1993) Behavior networks and force fields for simulating spinal reflex behaviors of the frog. In *Second Intl. Conf. on the Simulation of Adaptive Behavior* (ed. J-A. Meyer, H.L. Roitblat, and S.W. Wilson) pp. 172–181. MIT Press, Cambridge, MA.

Giszter, S.F., Kargo, W., and Davies, N.R. (1995) Dynamics of force field primitives in the reflex movements of spinal frogs. *Soc. Neurosci. Abstr.* **21** (abstract only).

Giszter, S.F., McIntyre, J., and Bizzi, E. (1989) Kinematic strategies and sensorimotor transformations in the wiping movements of frogs. *J. Neurophysiol.* **62** 750–767.

Giszter, S.F., Mussa-Ivaldi, F.A., and Bizzi, E. (1991) Equilibrium point mechanisms in the spinal frog. In *Visual Structures and Integrated Functions* (ed. M.A. Arbib and J.P. Ewert), pp. 223–237. Springer-Verlag, New York.

Giszter, S.F., Mussa-Ivaldi, F.A., and Bizzi E. (1993) Convergent force fields organized in the frog spinal cord. *J. Neurosci.* **13(2)** 467–491.

Graham, D. (1985) Pattern and control of walking in insects. *Adv. Insect Physiol.* **18** 31–140.

Haken, H. (1983) *Synergetics.* Springer-Verlag, Berlin.

Hallam, B.E. (1993) Fast robot learning using a biological model. International Workshop on Mechatronic Computer Systems for Perception and Action, Halmsted. Research Paper #630, Department of Artificial Intelligence, University of Edinburgh.

Hallam, B.E., Halperin J.R.P., and Hallam J.C.T. (1994) An ethological model for implementation in mobile robots. *Adapt. Behav.* **3** 51–79.

Halliday, T.R. (1974) The sexual behaviour of the smooth newt, *Triturus vulgaris* (Urodela, Salamandridae). *J. Herpetol.* 8 277–292.

Halliday, T.R. (1977) The effects of experimental manipulation of breathing behaviour on the sexual behaviour of the smooth newt *Triturus vulgaris. Anim. Behav.* **25** 39–45.

Halperin, J.R.P. (1991) Machine motivation. In *From Animals to Animats, Proc. Conf. on Simulation of Adaptive Behavior* (ed. J-A. Meyer and H. Roitblat) pp. 213–221. MIT Press, Cambridge, MA.

Halperin, J.R.P. (1995) Cognition and emotion in animals and machines. In *Comparative Approaches to Cognitive Science* (ed. H. Roitblat and J-A. Meyer) pp. 465–500. MIT Press, Cambridge, MA.

Halperin, J.R.P. and Dunham, D.W. (1992) Postponed conditioning: testing a hypothesis about synaptic strengthening. *Adapt. Behav.* **1** 39–63.

Halperin, J.R.P. and Dunham, D.W. (1993) Increased aggression after brief social isolation of adult fish: a connectionist model which organizes this literature. *Behav. Proc.* **28** 123–144.

Halperin, J.R.P., Dunham, D.W., and Ye, S. (1992) Social isolation increases social display after priming in *Betta splendens* but decreases aggressive readiness. *Behav. Proc.* **28** 13–32.

Halperin, J.R.P., Giri, Y., and Dunham, D.W. (1997) Different aggressive behaviors are exaggerated by facing vs. broadside subliminal stimuli shown to socially isolated Siamese fighting fish, *Betta splendens. Behav. Proc.* **40** 1–11.

Hansell, M.H. (1984) *Animal Architecture and Building Behaviour.* Longman, London.

Harrison, R. and Koch, C. (1998) An analog VLSI model of the fly elementary motion detector. In *Advances in Neural Information Processing Systems 10*, (ed. M.I. Jordan, M.J. Kearns, and S.A. Solla) pp. 880–886. MIT Press, Cambridge, MA.

Harvey, I., Husbands, P., and Cliff, D. (1994) Seeing the light: Artificial evolution, real vision. In *From Animals to Animats 3: Proc. Third Int. Conf. on Simulation of Adaptive Behavior* (ed. D. Cliff, P. Husbands, J-A. Meyer, and S.W. Wilson) pp. 392–401. MIT Press, Cambridge, MA.

Hebb, D. O. (1949) *The Organization of Behavior*. Wiley, New York.

Heiligenberg, W. (1969) The effect of stimulus chirps on cricket's chirping (*Acheta domesticus*). *Z. vergl. Physiol.* **65** 70–97.

Heiligenberg, W. (1974) A stochastic analysis of fish behaviour. In *Motivational Control Systems Analysis* (ed. D.J. McFarland) pp. 87–118. Academic Press, London.

Helmholtz, H. von (1867) *Handbuch der Physiologischen Optik*. Voss, Leipzig.

Henderson, J.V. (1999) Adaptive responses of herding animals to a robot vehicle. Ph.D. Thesis, University of Bristol.

Hendriks-Jansen, H. (1996) *Catching Ourselves in the Act*. MIT Press, Cambridge, MA.

Herman, R.M., Grillner, S., Stein, P.S.G., and Stuart, D.G. (1976) *Neural Control of Locomotion*. Plenum, New York.

Hinde, R.A. (1960) Energy models of motivation. *Symp. Soc. exp. Biol.* **14** 199–213.

Hinde, R.A. (1970) *Animal Behaviour*. 2nd edition. McGraw-Hill Book Company, New York.

Holland O.E. (1996) Grey Walter: the pioneer of real artificial life. In *Artificial Life V* (ed. C.G. Langton and K. Shimohara) pp. 34–41. MIT Press, Cambridge, MA.

Holland, O. and Melhuish, C. (1996) Some adaptive movements of animats with single symmetrical sensors. In *From Animals to Animats 4* (ed. P. Maes, M.J. Mataric, J-A. Meyer, J. Pollock, and S.W. Wilson) pp. 55–64. MIT Press, Cambridge, MA.

Holland, O. and Melhuish, C. (1999) Stigmergy, self-organization, and sorting in collective robotics. *Artif. Life* **5:2** 173–202.

Holland, P.C. (1977) Conditioned stimulus as a determinant of the form of the Pavlovian conditioned response. *J. Exp. Psychol.: Anim. Behav. Proc.* **3** 77–104.

Holland, P.C. and Straub, J.J. (1979) Differential effects of two ways of devaluing the unconditioned stimulus after Pavlovian appetitive conditioning. *J. Exp. Psychol.: Anim. Behav. Proc.* **5** 65–78.

Hölldobler, B. and Wilson, E.O. (1990) *The Ants*. Harvard University Press, Cambridge, MA.

Holst, E. von (1939) Entwurf eines Systems der Lokomotorischen Periodenbildungen bei Fischen. *Z. Verg. Physiol.* **26** 481–528.

Holst, E. v (1943) Uber relative Koordination bei Arthropoden. *Pflugers Archiv.* **246** 847–865.

Holst, E. von (1954) Relations between the central nervous system and the peripheral organs. *Br. J. Anim. Behav.* **2** 89–94.

Holst, von. E. (1973) *The Behavioral Physiology of Animals and Man.* University of Miami Press, Coral Gables, and Methuen and Co. Ltd., London.

Holst, E. von and Mittelstaedt, H. (1950) Das Reafferenzprinzip. *Naturwissenschaften* **37** 464–476.

Hopfield, J. (1982) Neural networks and physical systems with emergent collective and computational properties. *Proc. Natl. Acad. Sci. USA* **79** 2554–2558

Horridge G.A. (1966) *Symp. Soc. Exp. Biol.* **20**

Horsman, G. and Huber, F. (1994) Sound in crickets II. Modelling the role of a simple neural network in the prethoracic ganglion. *J. Comp. Physiol. A* **175** 399–413.

Houston, A.I. and McFarland, D.J. (1976) On the measurement of motivational variables. *Anim. Behav.* **24** 459–475.

Houston, A.I., Halliday, T.R. and McFarland, D.J. (1977) Towards a model of the courtship of the smooth newt *Triturus vulgaris*, with special emphasis on problems of observability in the simulation of behaviour. *Med. Biol. Eng. Comput.* **15** 49–61.

Howard, I.P. and Templeton, W.B. (1966) *Human Spatial Orientation.* Wiley, London.

Hoyle, G. (ed). (1976) *Identified Neurones and Behavior of Arthropods.* Plenum, New York.

Huber, F. (1983) Neural correlates of Orthopteran and Cicada phonotaxis. In *Neuroethology and Behavioural Physiology* (ed. F. Huber and H. Markle) Springer-Verlag, Berlin.

Hull, C.L. (1943). *Principles of behavior.* Appleton, New York.

Humphrys, M. (1996). Action selection methods using reinforcement learning. In *From Animals to Animats 4: Proc. SAB96* (ed. P. Maes, M.J. Mataric, J-A. Meyer, J. Pollock, and S.W. Wilson) pp. 135–144. MIT Press, Cambridge, MA.

Husbands, P. and Harvey, I. (1992) Evolution versus design: Controlling autonomous robots. In *Integrating Perception, Planning, and Action: Proc. 3rd Ann. Conf. On Artificial Intelligence, Simulation, and Planning.* pp. 139–146. IEEE Press.

Husbands, P., Harvey, I., and Cliff, D. (1993) An evolutionary approach to situated AI. In *Proc. 9th Bi-annual Conf. Of the Society for the Study of Artificial Intelligence and the Simulation of Adaptive Behaviour* (ed. A. Sloman) pp. 61–70. IOS Press.

Inhelder, B. and Piaget, J. (1964) *The Early Growth of Logic in the Child.* Routledge & Kegan Paul, London.

Iwamoto, N., Sugaya, R., and Sugi, N. (1990) Force-velocity relation of the frog skeletal muscle fibres shortening under continuously changing load. *J. Physiol.* **422** 185–202.

Jakobi, N., Husbands, P., and Harvey, I., (1995) Noise and the reality gap: The use of simulation in evolutionary robotics. In Advances in Artificial Life: Proc. 3rd Eur. Conf. On Artif. Life (ed. F. Moran, A. Moreno, J.J. Merelo, and P. Chacon) pp. 704–720. Springer Verlag, Berlin Heidelberg.

Jones, F.R.H. (1955) Photo-kinesis in the ammocoete larva of the brook lamprey. *J. exp. Biol.* **32** 492–503.

Johnson, D.W., Lovell III, G.H., and Murray J.J. (1997) Development of a coordinated motion controller for a front shovel excavator. In *Proc. American Nuclear Society 7th Topical Meeting on Robotics and Remote Systems.* pp 239–246. ANS, La Grange Park, IL

Kaelbling, L.P. (1991) An adaptable mobile robot. In *Toward a Practice of Autonomous Systems* (ed. F.J. Varela and P. Bourgine) pp. 41–47. MIT-Bradford, Cambridge, MA.

Kalman, R.E. (1963) Mathematical description of linear dynamical systems. *J. Soc. ind appl. Matt. (ser. A, Control)* **1** 152–192.

Kennedy, D. and Davis, W.J. (1977) Organization of invertebrate motor systems. In *Handbook of Physiology* (ed. R. Geiger, E. Kandel, and J.M. Brookhart) pp. 1023–1087. American Physiology Society, Bethesda, MD.

Kennedy, J.S. (1945) Classification and nomenclature of animal behaviour. *Nature* **156** 754.

Koza, J. (1991) Evolution of subsumption using genetic programming. In *Toward a Practice of Autonomous Systems* (ed. F.J. Varela and P. Bourgine) pp. 110–119. MIT-Bradford, Cambridge, MA.

Krieckhaus, E. and Wolf, G. (1968). Acquisition of sodium by rats: Interaction of innate mechanisms and latent learning. *J. of Comp. Physiol. Psychol.* **65** 197–201.

Kupferman, I. and Weiss, K.R. (1978) The command neurone concept. *Behav. Brain Sci.* **1** 3–39.

Kuwana, Y., Shimoyama, I., Sayama, Y., and Miura, H. (1996) A robot that behaves like a silkworm moth in the pheromone stream. In *Artificial Life V* (ed. C.G. Langton and K. Shimohara) pp. 370–376. MIT Press, Cambridge, MA.

Lambrinos, D., Maris, M., Kobayashi, H., Labhart, T., Pfeifer, R., and Wehner, R. (1997) An autonomous agent navigating with a polarized light compass. *Adapt. Behav.* **6 (1)** 131–161.

Lambrinos, D., Möller, R., Labhart, T., Pfeifer, R., and Wehner, R. (2000) A mobile robot employing insect strategies for navigation. *Rob. Auton. Syst.* **30** 39–64.

Lanz, P. and McFarland, D. (1995) On representation, goals and cognition. *Int. studies phil. Sci.* **9** 121–133.

Laverack, M.S. (1962) Responses of cuticular sense organs of the lobster, (*Homarus vulgaris*) (Crustacea). 1. Hair-peg organs as water current receptors. *Comp. Biochem. Physiol.* **5** 319–325.

Lipson, H. and Pollack, J.B. (2000) Automatic design and manufacture of robotic lifeforms. *Nature* **406** 974–978.

Lloyd, I.H. (1974) Stochastic identification methods. In *Motivational Control Systems Analysis* (ed. D.J. McFarland) pp. 169–207. Academic Press, London.

Loeb, J. (1918) *Forced Movements, Tropisms and Animal Conduct.* Lippincott, Philadelphia.

Lorenz, K. (1937) Uber die Bildung des Instinktbegriffes. *Naturwissenschaften* **25** 289–300, 307–318, 324–331.

Lorenz, K. (1950) The comparative method in studying innate behaviour patterns. *Symp. Soc. exp. Biol.* **4** 221–268.

Ludlow, A.R. (1980) The evolution and simulation of a decision maker. In *Analysis of Motivational Processes* (ed. F.M. Toates and T.R. Halliday). Academic Press, London.

Lund, H.H., Webb, B., and Hallam, J. (1997) A robot attracted to the cricket species *Gryllus bimaculatus*. In *Proc. Fourth European Conf. on Artificial Life* (ed. P.Husbands and I. Harvey) pp. 246–255. MIT Press, Cambridge, MA.

Mackintosh, N. J. (1974) The Psychology of Animal Learning. Academic Press: London.

Maes, P. (1989) The dynamics of action selection. *Proc IJCAI -89 Conf Detroit*

Maes, P. (1991a) Situated agents can have goals. In *Designing Autonomous Agents* (ed. P. Maes) pp. 49–70. MIT/Elsevier.

Maes, P. (1991b) A bottom-up mechanism for behavior selection in an artificial creature. In *From Animals to Animats: Proc. First Intl. Conf. on Simulation of Adaptive Behavior* (ed. J-A. Meyer and S.W. Wilson) pp. 238–246. MIT Press, Cambridge, MA.

Maes, P. (1991c) Learning behaviour networks from experience. In *Toward a Practice of Autonomous Systems*, (ed. F.J. Varela and P. Bourgine) pp. 48–57. MIT-Bradford, Cambridge, MA.

Maris, M. and te Boekhorst, R. (1996) Exploiting physical constraints: heap formation through behavioural error in a group of robots. In *Proc. IROS '96*, Osaka, Japan, pp. 1655–1660.

Marr, D. (1982) *Vision.* Freeman, San Francisco.

Martin, J.R., Cooper, S.E., and Ghez, C. (1995) Kinematic analysis of reaching in the cat. *Experimental Brain Research.* **102** 379–392.

Mataric M.J. (1990) Navigating with a rat brain: a neurobiologically inspired model for robot spatial navigation. In *From Animals to Animats: Proc. First Intl.*

Conf. on Simulation of Adaptive Behavior (ed. J-A. Meyer and S.W. Wilson) pp. 169–175. MIT Press, Cambridge, MA.

McFarland, D.J. (1965) Control theory applied to the control of drinking in the Barbary dove. *Anim. Behav.* **13** 478–492.

McFarland, D.J. (1971) *Feedback Mechanisms in Animal Behaviour.* Academic Press, London.

McFarland, D J (1974a) *Motivational Control Systems Analysis.* Academic Press, London.

McFarland, D.J. (1974b) Time-sharing as a behavioral phenomenon. In *Advances in The Study of Behavior* (ed. D.S. Lehrman, J.S. Rosenblatt, R.A. Hinde and E. Shaw). Academic Press, New York.

McFarland, D.J. (1977) Decision-making in animals. *Nature* **269** 15–21.

McFarland D. (1999) *Animal Behaviour.* Third Edition. Addison Wesley Longman, London.

McFarland, D. and Bösser, T. (1993) *Intelligent Behaviour in Animals and Robots.* MIT Press, Cambridge, MA.

McFarland, D.J. and Budgell, P. (1970) The thermoregulatory role of feather movements in the Barbary dove (*Streptopelia risoria*). *Physiol. Behav.* **5** 763–771.

McFarland, D.J. and Houston, A. (1981) *Quantitative Ethology: the State Space Approach.* Pitman, London.

McFarland, D.J. and Lloyd, I. (1973) Time-shared feeding and drinking. *Quart. J. Exp. Psychol.* **25** 48–61.

McFarland, D.J. and Rolls, B.J. (1972) Suppression of feeding by intracranial injections of angiotensin. *Nature* **236** 172–173.

McFarland, D.J. and Sibly, R. M. (1972) 'Unitary drives' revisited. *Anim. Behav.* **20** 548–563.

McFarland, D. and Sibly, R. (1975). The behavioural final common path. *Phil. Trans. Roy. Soc. B* **270** 265–293.

McFarland, D. and Spier, E. (1997). Basic cycles, utility and opportunism in self-sufficient robots. *Robotics and Auton. Syst.* **20** 179–190.

McGonigle, B. O. (1987) Non-verbal thinking by animals. *Nature* **325** 110–112.

McGonigle, B. (1990) Incrementing intelligent systems by design. In *From Animals to Animats: Proc. First Intl. Conf. on Simulation of Adaptive Behavior* (ed. J-A. Meyer and S.W. Wilson). MIT Press, Cambridge, MA.

McGonigle, B. (1999) Spatial representation as cause and effect: circular causality comes to cognition. In *Spatial Schemas and Abstract Thought* (ed. M. Gattis). MIT Press, Cambridge, MA.

McGonigle, B.O. and Chalmers, M. (1980) On the genesis of relational terms: A comparative study of monkeys and human children. *Antropol. Contemp.* **3** 236.

McGonigle, B.O. and Chalmers, M. (1984) The selective impact of question form and input mode on the symbolic distance effect in children. *J. Exp. Child Psychol.* **37** 525–554.

McGonigle, B. and Chalmers, M. (1986) Representations and strategies during inference. In *Reasoning and Discourse Processes* (ed. T. Myers). Academic Press, London.

McGonigle, B. and Chalmers, M. (1996) The ontology of order. In *Critical Readings on Piaget* (ed. L. Smith). Routledge, London.

McGonigle, B. and Chalmers, M. (1998) Rationality as optimized cognitive self-regulation. In *Rational Models of Cognition* (ed. M. Oaksford and N. Chater). Oxford University Press, Oxford.

McGonigle, B. and St Johnston, B. (1995) Applying staged cognitive growth principles to the design of autonomous robots. Technical paper, Laboratory for Cognitive Neuroscience, University of Edinburgh. Video demonstrated at JCI Exhibition, Royal College of Pathologists, London, 1995.

Melhuish, C., Holland, O.E., and Hoddell, S.E.J. (1998) Collective sorting and segregation in robots with minimal sensing. In *From Animals to Animats 5* (ed. R. Pfeifer, B. Blumberg, J-A. Meyer, and S.W. Wilson). MIT Press, Cambridge, MA.

Metz, H.A.J. (1974) Stochastic models for the temporal fine structure of behaviour sequences. In *Motivational Control Systems Analysis* (ed. D.J. McFarland) pp. 5–86. Academic Press, London.

Moiseff, A., Pollack, G.S., and Hoy, R.R. (1978) Steering responses of flying crickets to sound and ultrasound: mate attraction and predator avoidance. *Proc. Nat. Acad. Sci. USA*, **75** 4052–4056.

Mondada, F., Franzi, E., and Ienne, P. (1993) Mobile robot miniaturization: a tool for investigation in control algorithms. *Proc. 3^d Int. Symp. On Experimental Robotics ISER-93* pp. 501–513

Monismith, S.G., Koseff, J.R., Thompson, J., O'Riordan, C., and Nepf, H. (1990) A study of bivalve siphonal currents. *Limnol. Oceanogr.* **35(3)** 680–696.

Moore, P.A,. and Atema, J. (1991) Spatial information in the three-dimensional fine structure of an aquatic odor plume. *Biol. Bull.* **181** 408–418.

Moore, P.A., Scholtz, N. and Atema, J. (1991) Chemical orientation in American lobsters, *Homarus americanus*, in turbulent odor plumes. *J. Chem. Ecol.* **17** 1293–1307.

Morgan, C.L. (1894) Introduction to comparative psychology. Scott, London.

Morse, T.M., Ferrée, T.C., and Lockery, S.R. (1998) Robust spatial navigation in a robot inspired by *C. elegans*. *Adapt. Behav.* **6** 391–408.

Mottet, D., Bootman, R.J., Guiard, Y., and Laurent, N. (1994) Fitts' Law in two-dimensional task space. *Exp. Brain Res.* **100** 144–148.

Moyer, R.S. (1973) Comparing objects in memory: Evidence suggesting an internal psychophysics. *Percept. Psychophys.* **13** 180–184.

Müller-Wilm, U., Dean, J., Cruse, H., Weidermann, H.J., Eltze, J., and Pfeiffer, F. (1992) Kinematic model of the stick insect as an example of a 6-legged walking system. *Adapt. Behav.* **1** 155–169.

Mussa-Ivaldi, F.A. (1992) From basis functions to basis fields: Using vector primitives to capture vector patterns. *Biol. Cybern.* **67** 479–489.

Mussa-Ivaldi, F.A. and Gandolfo, F. (1993) Networks that approximate vector-valued mappings. Proc. 1993 IEEE Conf. Neural Networks, San Francisco, CA. Pp. 1973–1978.

Mussa-Ivaldi, F.A. and Giszter, S.F. (1992) Vector field approximation: a computational paradigm for motor control and learning. *Biol. Cybern.* **67** 491–500.

Nehmzow, U. and McGonigle, B. (1994), Achieving rapid adaptations in robots by means of external tuition. In *From Animals to Animats 3: Proc. Third Int. Conf. on Simulation of Adaptive Behavior* (ed. D. Cliff, P. Husbands, J-A. Meyer, and S.W. Wilson) pp. 301–308. MIT Press, Cambridge, MA.

Nicolis, G. and Prigogine, I. (1977) *Self-Organization in Non-Equilibrium Systems.* Wiley and Sons, New York.

Nolfi, S., Floreano, D., Miglino, O., and Mondada, F. (1994) How to evolve autonomous robots: Different approaches in evolutionary robotics. In *Artificial Life IV* (ed. R.A. Brooks and P. Maes) pp. 190–197. MIT Press, Cambridge, MA.

Oatley, K. and Toates, F.M. (1971) Frequency analysis of the thirst control system. *Nature* **232** 562–564.

Oldfield, B.P. (1980) Accuracy of orientation in female crickets, *Teleogryllus oceanus* (Gryllidae): dependence on song spectrum. *J. Comp. Physiol.* **141** 93–99.

Paivio, A. (1975) Perceptual comparisons through the mind's eye. *Memory cogn.* **3** 635–647.

Pearson, K.G. 1972) Central programming and reflex control of walking in the cockroach. *J. Exp. Biol.* **56** 173–193.

Pfann, K.D., Hoffman, D.S., Gottlieb, G.L., Strick, P.L., and Corcos, D.N. (1998) Common principles underlying the control of rapid, single degree-of-freedom movements at different joints. *Exp. Brain Res.* **118** 35–51.

Pinker, S. (1997) *How the Mind Works.* W.W. Norton, New York.

Pinsker, H.M. and Ayers, J. (1983) Neuronal Oscillators. In *The Clinical Neurosciences. Section Five. Neurobiology, Chapter 9.* W.D. Willis.

Pollack, G.S. (1986) Discrimination of calling song models by the cricket, *Teleogryllus oceanus*: the influence of sound direction on neural encoding of the stimulus temporal pattern and on phonotactic behaviour. *J. Comp. Physiol. A* **158** 549–561.

Popov, A.V. and Shuvalov, V.F. (1977) Phonotactic behaviour of crickets. *J. Comp. Physiol.* **119** 111–126.

Prescott, T.J. and Ibbotson, C. (1997) A robot trace-maker: modeling the fossil evidence of early invertebrate behaviour. *Artificial Life* **3** 289–306.

Prescott, T.J., Redgrave, P., and Gurney, K. (1999) Layered control architectures in robots and vertebrates. *Adapt. Behav.* **7** 99–127.

Prosser, C.L. (1973) *Comparative Animal Physiology*. Third edition. W.B. Saunders Co., Philadelphia.

Pylyshyn, Z. (1984) *Computation and Cognition: Toward a Foundation for Cognitive Science*. MIT Press, Cambridge, MA.

Quartz, S. and Sejnowski T. (1997) The neural basis of cognitive development: a constructivist manifesto. *Behav. Brain Sci.* **20(4)** 537–596.

Ramachandran, V.S. and Blakeslee S. (1998) *Phantoms in the Brain: Human Nature and the Architecture of the Mind*. Fourth Estate, London.

Rescorla, R.A. and Wagner, A.R. (1972). A theory of Pavlovian conditioning: Variations in the effectiveness of reinforcement and nonreinforcement. In *Classical Conditioning II: Current Theory and Research* (ed. A.H. Black and W.F. Prokasy) pp. 64–99. Appleton-Century-Crofts, New York.

Robinson, G.E. (1992) Regulation of division of labour in insect societies. *Ann. Rev. of Entomol.* **37** 637–665.

Robotics and Autonomous Systems (2000) Special Issue on Biomimetic Robots. (Ed. C. Chang and P. Gaudiano) *Rob. Auton. Syst.* **30 (1–2)**.

Ryan, M.J. and Keddy-Hector, A. (1992) Directional patterns of female mate choice and the role of sensory biases. *Am. Nat.* **139** S4–S35.

Sabes, P.N., Jordon, N.I., and Wolpert, D.N. (1998) The role of inertial sensitivity in motor planning. *J. Neurosci.* **18** 5948–5957.

Sandeman, D.C. and Atwood, H.L. (1982) *The Biology of Crustacea, Vol. 4. Neural Integration and Behavior*. Academic Press, New York.

Santschi, F. (1911) Observations et remarques critiques sur les mechanisms de l'orientation. *Rev. Suisse Zool.* **19** 303–338.

Schildberger, K. (1984) Temporal selectivity of identified auditory interneurones in the cricket brain. *J. Comp. Physiol.* **155** 171–185.

Schmitz, J. (1985) Control of the leg joints in stick insects: differences in the reflex properties between the standing and the walking states. In *Insect Locomotion* (ed. M. Gewecke and G. Wendler, G.) pp. 27–32. Parey, Hamburg, Berlin.

Schöne, H. (1984) Spatial Orientation. Princeton University Press, NJ.

Segejeva, M.V. and Popov, A.V. (1994) Ontogeny of positive phonotaxis in female crickets, *Gryllus bimaculatus* de Greer: dynamics of sensitivity, frequency-intensity domain and selectivity to the temporal pattern of male calling song. *J. Comp. Physiol. A* **174** 381–389.

Selverston, A.I. and Moulins M. (1987) *The Crustacean Stomatogastric System*. Springer-Verlag, Berlin.

Shadmehr, R. (1991) *A Computational Theory for Control of Posture and Movement in a Multi-joint Limb.* Ph.D. dissertation, Technical Report 91-07, Center for Neural Engineering, University of Southern California, Los Angeles.

Shanks, D. (1995). *The Psychology of Associative Learning.* Cambridge University Press, Cambridge.

Sherrington, C. S. (1906) *The Integrative Action of the Nervous System.* Charles Scribner and Sons, New York.

Sherrington, C.S. (1918) Observations on the sensual role of the proprioceptive nerve-supply of the extrinsic ocular muscles. *Brain* **41** 332–343.

Shipley, B.E. and Colwill, R.M. (1996). Direct effects on instrumental performance of outcome revaluation by drive shifts. *Anim. Learn. Behav.* **24** 57–67.

Sibly, R.M. and McFarland, D.J. (1976) On the fitness of behavior sequences. *Am. Nat.* **110** 601–617.

Simpson, M.J.A. (1968) The display of the Siamese fighting fish, (*Betta splendens*). *Anim. Behav. Monog.* **1**.

Sims, K. (1994) Evolving 3D morphology and behavior by competition. In *Artificial Life IV* (ed. R.A. Brooks and P. Maes) pp. 28–39. MIT Press, Cambridge, MA.

Skinner, B.F. (1938) *The Behavior of Organisms.* Appleton-Century-Crofts, New York.

Skinner, B.F. (1958) Reinforcement today. *Am. Psychol.* **13** 94–99.

Snaith, M.A. and Holland, O.E. (1990) An investigation of two mediation strategies suitable for behavioural control in animals and animats. In *From Animals to Animats: Proc. First Intl. Conf. on Simulation of Adaptive Behavior* (ed. J-A. Meyer and S.W. Wilson) pp. 255–262. MIT Press, Cambridge, MA.

Sparks, J. (1982) *The Discovery of Animal Behaviour.* Collins Sons and Co. Ltd., London.

Spier, E. and McFarland, D. (1996). A finer-grained motivational model of behaviour sequencing. In *From Animals to Animats 4: Proc. SAB96.* (ed. P. Maes, M.J. Mataric, J-A. Meyer, J. Pollack, and S.W. Wilson) pp. 255–263. MIT Press, Cambridge, MA.

Spier, E. and McFarland, D. (1997). Possibly optimal decision making under self-sufficiency and autonomy. *J. Theor. Biol.* **189** 317–331.

Spier, E. and McFarland, D. (1998). Learning to do without cognition. In *From Animals to Animats 5: Proc. SAB98* (ed R. Pfeifer, B. Blumberg, J-A. Meyer, and S.W. Wilson) pp. 38–47. MIT Press, Cambridge, MA.

Stabel, J., Wendler, G., and Scharstein, H. (1989) Cricket phonotaxis: localization depends on recognition of the calling song pattern. *J. Comp. Physiol.* **165** 165–177.

Stark, L. (1959) Stability, oscillations and noise in the human pupil servomechanism. *Proc. I.R.E.* **47** 1925–1936.

Steels, L. (1993) Building agents with autonomous behaviour systems. In *The Artificial Life Route to Artificial Intelligence: Building Situated Embodied Agents* (ed. L. Steels and R. Brooks) Lawrence Erlbaum Associates, New Haven, CT.
Steels, L. (1994) The artificial life roots of artificial intelligence. *Artif. Life* **1** 89–125.
Stein, P.S.G. (1976) Mechanisms for interlimb phase control. In *Neural Control of Locomotion* (ed. R.M. Herman, S. Grillner, P.S.G. Stein, and D.G. Stuart) pp. 465–487. Plenum, New York.
Stein, P.S.G. (1978) Motor systems, with specific reference to the control of locomotion. *Ann. Rev. Neurosci.* **1** 61–81.
Stein, P.S.G., Martin, L.I., and Robertson, G.A. (1986) The forms of a task and their blends. In *Neurobiology of Vertebrate Locomotion* (ed. P.S.G. Stein, D.G. Stuart, H. Forssberg, and R.M. Herman) pp. 201–216. Macmillan, London.
Stephens, D.W. and Krebs, J.R. (1986) *Foraging Theory*. Princeton University Press, Princeton, NJ.
Stout, J.F., DeHaan, C.H., and McGhee, R. (1983) Attractiveness of the male *Acheta domestica* calling song to females, I: Dependence on each of the calling song features. *J. Comp. Physiol. A* **153** 509–521.
Stout, J.F. and McGhee, R. (1988) Attractiveness of the male *Acheta domestica* calling song to females, II: The relative importance of syllable period, intensity and chirp rate. *J. Comp. Physiol.* **164** 277–287.
Sutton, R. and Barto, A. (1981a). An adaptive network that constructs and uses an internal model of its world. *Cogn. Brain Theory* **4** 217–246.
Sutton, R. and Barto, A. (1981b). Toward a modern theory of adaptive networks: Expectation and prediction. *Psychol. Rev.* **88** 135–170.
Terrace, H.S. and McGonigle, B.O. (1994), Memory and representation of serial order by children, monkeys and pigeons, *Curr. Directions in Psychol. Sci.* **3:6** 180–185.
Thorndike, E.L. (1898) Animal Intelligence: an experimental study of the associative processes in animals. *Psychol. Rev. Monogr. Suppl.* **2:8** 1–16 .
Thorndike, E.L. (1911) *Animal Intelligence*. Macmillan, New York.
Thorndike, E.L. (1913) *The Psychology of Learning. (Educational Psychology II)*. Teachers College, New York.
Thorson, J., Weber, T., and Huber, F., (1982) Auditory behaviour of the cricket II. Simplicity of calling-song recognition in *Gryllus* and anomalous phonotaxis at abnormal carrier frequencies. *J. Comp. Physiol. A* **146** 361–378.
Tinbergen, N. (1950) The hierarchical organization of nervous mechanisms underlying instinctive behaviour. *Symp. Soc. Exp. Biol.* IV 305–312.
Tinbergen, N. (1951) *The Study of Instinct*. Oxford University Press, Oxford, New York, and London.
Tinbergen, N. (1953) *The Herring Gull's World*. Collins, London.

Toates, F.M. (1975) *Control Theory in Biology and Experimental Psychology.* Hutchinson Educational, London.

Toates, F.M. (1980) *Animal Behaviour - a Systems Approach.* Wiley & Sons, Chichester.

Tolman, E.C. (1932). *Purposive Behavior in Animals and Men.* Appleton-Century-Crofts, New York.

Tovish, A. (1982) Learning to improve the availability and accessibility of resources. In *Functional Ontogeny* (ed. D. McFarland). Pitman, London.

Trabasso, T. and Riley, C.A. (1975) On the construction and use of representations involving linear order. In *Information Processing and Cognition: The Loyola Symposium* (ed. R.L. Solso). Erlbaum, Hillsdale, NJ.

Trivers, R. (1985) *Social Evolution.* Benjamin-Cummings, Menlo Park, CA.

Ullyott, P. (1936) The behaviour of *Dendrocoelum lacteum*: I and II *J. Exp. Biol.* **13** 253–264, 265–278.

Vaughan, R., Sumpter, N., Frost, A., and Cameron, S. (1998) Robot sheepdog project achieves automatic animal control. In *From Animals to Animats 5: Proc. SAB98.* (ed R. Pfeifer, B. Blumberg, J-A. Meyer, and S.W. Wilson), MIT Press, Cambridge, MA.

Walter, W.G. (1953) *The Living Brain.* W.W.Norton, New York.

Watson, J.B. (1919) *Psychology from the Standpoint of a Behaviourist.* Lippincott, Philadelphia.

Weber, T. and Thorson, J. (1988) Auditory behaviour in the cricket IV. Interaction of direction of tracking with perceived temporal pattern in split-song paradigms. *J. Comp. Physiol.* **163** 13–22.

Wendler, G. (1964) Laufen und Stehen der Stabheuschrecke: Sinnesborsten in den Beingelenken als Glieder von Regelkereisen. *Z. vergl. Physiol.* **48** 198–250.

Wendler, G. (1966) Coordination of walking movements in arthropods. *Symp. Soc. Exp. Biol.* **20** 229–249.

Wendler, G. (1990) Pattern recognition and localization in cricket phonotaxis. In *Sensory Systems and Communication in Arthropods.* Birkhauser Verlag, Basel

Wiener, N. (1948) *Cybernetics: or Control and Communication in the Animal and the Machine.* Technology Press, Cambridge, MA.

Wiersma, C.A.G. and Roach, J.L.M. (1977) Principles in the organization of invertebrate sensory systems. In *Handbook of Physiology, Section 1: The Nervous System. Volume 1: Cellular Biology of Neurones* (ed. J.M. Brookhart, V.V.B. Mountcastle, and E.M. Kandel) pp. 1089–1087. Williams and Watkins Co., Baltimore.

Widrow, B. and Hoff, M.E. (1960). Adaptive switching circuits. In *1960 IRE WESCON Convention Record (pt. 4),* pp. 96–104.

Winfield, A.F.T and Holland, O.E. (2000). The application of wireless local area network technology to the control of mobile robots. *Microproc. Microsyst.* **23:10** 597–607.

Wohlers, D.W. and Huber, F. (1980) Processing of sound signals by six types of neurones in the prothoracic ganglion of the cricket, *Gryllus campestris. J. Comp. Physiol.* **146** 161–173.

Index

action potential 42
action selection 219–21
activation in Maes networks 36–7, 106–7, 109–11
adaptation
 in model neurones 165
 in receptors 95
after-discharge
 in fighting fish 164, 176
 in machine motivation model 165, 170, 171
animal-like robots, evolution of 14-41
animal-robot hybrids 40
antennae
 lobster 147
 moth 40, 232, 234
antennules 57–8, 148
ants 45, 78, 80, 91
associationists 156, 159
auditory system of cricket 61–8, 75
autonomy 230–1

behaviour patterns in Grey Walter tortoise 25–7, 29
behavioural competence 34
behavioural primitives 104
Betta splendens,(Siamese fighting fish) 162–4, 176
biomimicry 48–51, 56, 231
blueprints for constructions 76
Braitenberg machines 35, 37
bricolage 18
building behaviour 76, 91
built systems 21–22

central control 119–20, 132, 133
central nervous system (CNS) 94, 100, 119, 141, 143
cerebellum 121–2
chemosensors 47
chemotaxis 49–50, 52
classical conditioning 158–9
closed loop 120
cognition 186, 188, 204–5, 223–4
command neurones 143, 144
communication 74, 162
compensation
 for tissue loss 53
 for disturbances 127–8
compensatory reflexes 152–3
competition
 in Maes networks 36
 in machine motivation model 167–8
computers and robots 32
concentration gradient, following 47–8, 52, 57–8
concurrency 106, 110–11
conditional response 158, 191
conditional stimulus 158, 183
control strategies
 in crustacea 132, 155
 in frog wiping reflexes 98, 101
control systems 16, 18, 33
 biologically inspired 38–9, 165, 177
 in insect walking 122, 126, 138–9
control systems theory 2, 8, 11
controller, adaptive 178, 180
co-ordination
 of activities in social insect colonies 76–8
 of frog limb movements 97, 99
 of lobster behaviour 141, 144–5
 of motor activities 119–22
 of stick insect walking 123, 132, 136
cortex 121–2
cybernetics 1, 6

decision making in animals 12–13
declarative knowledge 187, 190–1
declarative representations 205, 209
degrees of freedom 97, 104, 132–3, 139, 235
delta rule 195
designed systems 29, 22
drives 36, 160–1

eligibility trace 196
emergence 18, 29–31, 78
error signals 134, 135, 178
ethology 1, 14, 15, 162
ethology, artificial 41, 235
evolved systems 17, 20
exclusion in Maes networks 106–8, 110
executable states in Maes networks 36–7, 106–8
explicit knowledge 187–8, 190, 223
extinction 196–7
extrapyramidal system 121

feedback
  in cuttlefish 120
  in lobsters 141
  in machine motivation model 165, 168, 180
  in stick insect walking model 126, 133–8
feedforward
  control function 178–80
  neural networks 126, 128
filtering of signals 60, 66, 69, 132
Fitt's law 114, 116
fixed-action patterns 160–5, 167–8, 171, 176–7
fly optomotor reflex 5–6
fly vision system model 38, 234
force fields 100–106, 108, 110, 112–118
frequency selectivity in cricket phonotaxis 68, 72
functional models 12–13

gait 17, 123, 136, 141
generalization in learning 166–7, 178–80

goals of behaviour 36–7, 105, 149, 195, 221
gradients of stimulation 19, 46, 47–8, 57–8, 214–16
Grey Walter tortoise 24–32
grown systems 17, 19–22

hidden states 190
hierarchy 20, 155, 160, 214

ill-posed problems 98–100, 102–3
implicit representations 187, 188
impulse response 8
inclinometers 148
inflow theory 3
information processing 14, 223
inhibition
  in cricket phonotaxis model 67
  of eating by thirst 6
  in machine motivation model 165, 167, 170
  in Maes networks 37, 113
  in robot lobster control system 145, 155
  in stick insect walking model 132
  in subsumption architectures 34
  in telotaxis 45
innate releasing mechanism 160–1
instinct 159–60, 171
instrumental learning 181–4, 189, 194–5
intelligence, artificial (AI) 32–35, 223
intelligent behaviour in AI 33, 223
internal representation 33–4
interneurones 116, 120, 129, 152
IR (infra red) sensors 39, 80, 172, 211–12, 214

joint angles
  in frog 97, 99, 100
  in lobster 150
  in stick insect 126, 128, 136

kineses 43, 47, 153
klinokinesis 43
klinotaxis 44

Index 259

labelled lines 147
latency 64, 75, 148
lateral inhibition 155, 167
law of effect 182, 184
leaky integration 68
learning
 in biologically-inspired controllers 38–9
 in fixed-action patterns 164
 instrumental 181–6
 in machine motivation model 168, 171, 172, 174–6, 178–81
 in Maes networks 110–12
 in Pavlovian conditioning 157–8
 procedural 188, 189–204
 in simple robots 36
learning curve 200
lobster-inspired robots 47–58, 139–155
localization 43, 48, 68, 69, 75
Lorenz model of motivation 1, 160–1

Maes networks 106, 108, 112
magnet effect 1
mechanoreceptors 42, 130
menotaxis 45
microprocessors 21, 39, 61, 80, 214
modality 16, 53–4, 147
models in ethology 1–13
modularity
 of movement control 98, 102
 structural and functional 19, 20, 22–3
modulation 106, 123, 141, 148–50, 152
motivational state 11, 165, 167–8, 171, 195
motor-equivalence 98–9, 103
motor neurones 67–8, 96, 120, 132, 141
muscles 3–4, 93–7, 116–18, 119–122

navigation 35, 38, 214–15, 216
nervous system 19, 40, 93–4, 133, 174
Neural Network Adaptive Controller (NNAC) 166, 179–80
neural networks 38, 39, 40, 166, 178
neuronal networks 140, 144
neurones 42
 auditory neurones in crickets 63, 65–8
 in cerebellum 122
 in machine motivation model 165, 167–175
 in mechanisms underlying locomotion 141, 144, 145
 modelled in control systems 38–9
 modelled in stick insect walking controller 126, 130, 133
 and muscles 93
 in Neural Network Adaptive Controller 178–80
 pyramidal 121
 in stick insect 129
noise 16, 39, 79, 227
non-executable states in Maes networks 36, 107–8
non-linearity 8, 102, 132, 215

observability 12
open loop 120
operant conditioning 184
opportunism 192, 221
orientation 43–47
 auditory 59
 chemo-orientation 47–9
 chemo-orientation strategies in lobsters 54–58
 in hearing robot 214–16
 in lobsters 153
 relative to gravity 148
 in Siamese fighting-fish displays 163
 spatial 76–7
orthokinesis 43
outcome devaluation
 in animals 189–91
 procedural account of 193–5
 tested in simulation 197–204
outflow theory 3–4

parallel systems 8, 36
pattern generators 113, 118, 124–5, 126, 141
peripheral control 119–120
phonotaxis, cricket 63, 67–8, 75
phototaxis 43–4
planning 76, 77, 101–4, 112, 192
plasticity 111, 222
position error 178

postural control mechanisms 119, 121–2
posture
    in fighting fish 162–3, 171
    in frogs 98–103
    in lobsters 148, 149, 152
prepotent reflexes 27
primitives, force field and movement 101–118
priority, in behavioural architectures 34, 82, 155
procedural knowledge 187–8, 190, 223
proportional controller 134
proprioception
    in feedback 116, 117, 120, 141
    in reflexes 149, 152
pyramidal system 121

rationalism 156
reactive systems 59, 111–2, 153, 210, 214
reafference 4, 6
real time 39, 70, 211, 227, 229
receptive fields 147, 166, 172, 178–9
reciprocal inhibition 119
recognition 32, 75, 161, 164
recruitment of motor units 144–7
redundancy 102, 104, 125
reflexes
    conditional 157–8, 191
    co-ordinated 119–21
    frog spinal 97
    frog wiping 114–18
    in lobster 141, 149–50, 152–3
    optomotor 5
    prepotent 27
    simple muscular 96
    in stick insects 128, 133
reinforcement
    in classical conditioning 158
    in machine motivation model 171–2, 174–7
    in Maes networks 111
    in S-R conditioning 182–6, 196–204
relative co-ordination 1, 120, 122
releasers 149, 152, 161, 165, 172
replication of experiments 86, 229, 233
reverse engineering 141–2
rheotaxis 152
robot-animal hybrids 40

robotics
    behaviour-based 33–8, 210–11, 217, 223
robots
    for arm-wrestling frog 115–18
    Ben Hope 172–7
    biomimetic 38–40
    cricket 59-75
    Edinburgh R2 213–18
    Grey Walter tortoise 24–32
    hearing 212–213
    lobster biomimic 139–155
    Nomad 218–20
    Robolobster 48–58
    stick insect walking 138, 139
    TRex 177–81
    Ubots 177–92
robots in biology
    advances needed 233–5
    dangers 233
    desirability 227–8
    disadvantages 228–30
    important aspects 231–2
    indirect benefits 232
    methodologies 233
    necessity 226–7
    uses 230–1
robustness 34, 59, 80, 125, 230
rule sets 82, 84, 89

self-organization 77–8
self-organized clustering 82, 92
sensory-motor integration 99
sensory-motor modules 35
sequences of behaviour 14, 33, 76
    in behaviour-based robots 34
    in fighting fish 63–4
    in frog wiping reflexes 105–6, 111
    in Grey Walter tortoise 26, 27
    in lobsters 140, 144, 148–51
    in Maes networks 37, 113
    in newt courtship 9–11
Siamese fighting fish 162–4, 176
sign stimulus 160–1
simulation, computer
    advantages 229
    of clustering and sorting 79
    disadvantages 54–5

Index

of evolution 39–40
of force-field models 103–6, 108–10, 112–13
of insect walking 125, 126, 128, 134–9
as modelling tool 14–15
problems 226–9, 232
of procedural learning 197–204
as tool in mobile robotics 32
simulation, hardware 230
Skinner, B.F. 157, 184–5
Skinner box 184, 189–90, 196, 200
social insects 76–9, 91–2
somesthetic system 94
specific nerve energies 42
specific satiety 189, 195
spikes in neurones 60, 64, 68, 75
spinal cord 96, 97–102, 111–8, 120–1
stance phase in walking 122–3, 126, 130–3, 137, 143
state space 111, 192, 198, 203, 213
statocysts 147–8
stick insect walking models 122–139
stimulus-response (S-R) theory of learning 167–8, 170–2, 182–4, 186, 190–1
stochastic analysis 6, 8, 13
subsumption architecture 33–6, 218
swing phase in walking 122–3, 126–32, 143–5, 152
switching mechanisms
  in cricket behaviour 69
  in frog behaviour 98, 106, 109–10
  in robot behaviour 24, 35–6
  in stick insect behaviour 134
symbolic distance effect 206, 207, 209
symbolic representation 33, 210, 223

taxes 43, 153
telotaxis 45
temporal integration 49, 79
termites 77
thresholds
  in clustering and sorting robots 80–2
  in cricket auditory system 68–9, 72
  in fighting fish display 164–5
  in Grey Walter tortoise robot 25–7, 29–30, 37
  in lobster robot 58
  in neurons 42
  in newt behaviour 10
  in stick insect walking model 130
tracking
  odour 48–50, 56–8
  sound source 65, 69
trade-offs in biology 13, 57, 181, 203
transducers 42, 147, 234–5
transfer function 6, 12
transitive inference 205, 207, 209
tropotaxis 44, 52

unconditional response 158, 191
unconditional stimulus 158
unobservability 36

vector fields 104, 127
vector summation 102, 104–5, 112
vestibular system 40, 95, 122, 147
vision 3, 122
vision systems on robots 214, 216

walking 119–20
  in lobsters 139, 142–6
  in stick insects 122–139
world model 33
world as its own best model 35, 138